Lecture Notes in Computer Science 16308

Founding Editors

Gerhard Goos
Juris Hartmanis

Editorial Board Members

Elisa Bertino, *Purdue University, West Lafayette, IN, USA*
Wen Gao, *Peking University, Beijing, China*
Bernhard Steffen, *TU Dortmund University, Dortmund, Germany*
Moti Yung, *Columbia University, New York, NY, USA*

The series Lecture Notes in Computer Science (LNCS), including its subseries Lecture Notes in Artificial Intelligence (LNAI) and Lecture Notes in Bioinformatics (LNBI), has established itself as a medium for the publication of new developments in computer science and information technology research, teaching, and education.

LNCS enjoys close cooperation with the computer science R & D community, the series counts many renowned academics among its volume editors and paper authors, and collaborates with prestigious societies. Its mission is to serve this international community by providing an invaluable service, mainly focused on the publication of conference and workshop proceedings and postproceedings. LNCS commenced publication in 1973.

Siva Teja Kakileti · Geetha Manjunath ·
Robert G. Schwartz · Eddie Y. K. Ng
Editors

Artificial Intelligence over Infrared Images for Medical Applications

4th International Conference, AIIIMA 2025
Virtual Event, November 15, 2025
Proceedings

Editors
Siva Teja Kakileti
Niramai Health Analytix Pvt. Ltd.
Bengaluru, Karnataka, India

Robert G. Schwartz
Piedmont Physical Medicine
and Rehabilitation
Greenville, SC, USA

Geetha Manjunath
Niramai Health Analytix Pvt. Ltd.
Bengaluru, Karnataka, India

Eddie Y. K. Ng
Nanyang Technological University
Singapore, Singapore

ISSN 0302-9743　　　　　　　　ISSN 1611-3349 (electronic)
Lecture Notes in Computer Science
ISBN 978-3-032-10989-7　　　　ISBN 978-3-032-10990-3 (eBook)
https://doi.org/10.1007/978-3-032-10990-3

Preface

In recent years, infrared imaging hardware has seen significant advancements, accompanied by a surge in its application within the medical field. As a non-contact, non-invasive, and radiation-free modality, infrared imaging offers distinct advantages, including cost-effectiveness and portability, making it an attractive alternative to other imaging technologies. Additionally, infrared thermal images possess unique characteristics, such as inherent anonymity and the ability to be analyzed in both thermal and imaging feature spaces. These features have generated substantial interest across various research communities. The emergence of innovative startups developing breakthrough solutions in medical infrared thermology, combined with novel machine learning algorithms applied to thermal images, underscores the growing impact of this technology.

Building on the success of previous years, AIIIMA 2025 continued to serve as a dedicated forum for discussing this specific and evolving area of research. Our aim was to foster collaboration and innovation within the community, as we explore the potential of AI-driven infrared imaging to revolutionize medical diagnostics and treatment. This year, we invited submissions covering a wide range of topics, including but not limited to the applications of artificial intelligence in medical infrared imaging for cancer screening, cancer diagnosis, cancer risk assessment, treatment monitoring, sports injury, and pain management. AIIIMA 2025 also hosted a data challenge that involved open-sourcing datasets and libraries, providing new researchers with accessible resources to engage with and contribute to this field.

This year, a total of 23 papers were submitted for consideration. Following a rigorous peer-review process, 13 papers were accepted for publication. Each submission was evaluated by at least three external reviewers and one program chair in a single-blind review process. Reviewers were carefully selected for their expertise in medical imaging, particularly medical infrared imaging, and all submissions were assigned to avoid conflicts of interest. The Conference Management Toolkit (CMT) was used to manage the submission and review workflow efficiently. Authors received detailed feedback and constructive suggestions to help enhance the quality of their manuscripts.

AIIIMA 2025 was held as a fully virtual conference with no registration fees, ensuring broad accessibility. The conference program featured oral presentations of all the accepted papers. Additionally, to further inspire research and innovation in this field, the program also included a keynote lecture delivered by Prof. Prathosh A. P., Assistant Professor, Indian Institute of Science (IISc), Bengaluru, India.

We extend our sincere gratitude to the authors for submitting their manuscripts to the fourth edition of AIIIMA and for their diligent efforts in collaborating with reviewers to enhance the quality of their work. We also express our appreciation to the reviewers for their time and for providing detailed feedback and constructive suggestions to the authors within the given timeframe. Our heartfelt thanks go to the Program Chairs and Organizing Committee, whose leadership, vision, and tireless efforts made this conference possible. Finally, we would like to thank our sponsor, Niramai Health Analytix Pvt Ltd., for their invaluable support in publicizing and sponsoring the AIIIMA conference. Lastly, we would also like to thank LNCS for publishing the AIIIMA 2025 proceedings.

November 2025

Siva Teja Kakileti
Geetha Manjunath
Robert G. Schwartz
Eddie Y. K. Ng

Organization

Organizing Committee

Siva Teja Kakileti	Niramai Health Analytix Pvt Ltd., India
Geetha Manjunath	Niramai Health Analytix Pvt Ltd., India
Robert G. Schwartz	American Academy of Thermology, USA
Ng Yin Kwee	Nanyang Technological University, Singapore

Program Committee

Siva Teja Kakileti	Niramai Health Analytix Pvt Ltd., India
Geetha Manjunath	Niramai Health Analytix Pvt Ltd., India
Robert G. Schwartz	American Academy of Thermology, USA
Ng Yin Kwee	Nanyang Technological University, Singapore
Vaibhav Rajan	Google DeepMind, India

External Reviewers

Aishwarya Jayagopal	National University of Singapore, Singapore
Akash Narayana	International Institute of Information Technology, Bangalore, India
Bhanu Sekhar Guttikonda	New England College, USA
Marcos Leal Brioschi	Brazilian Medical Thermography Association, Brazil
Matheus Baffa	University of São Paulo, Brazil
Minerva Panda	Cornell University, USA
Mouad El Omari	Mohammed V University, Morocco
Mustafa Yurdakul	Kirikkale University, Turkey
Ninad Aithal	Niramai Health Analytix Pvt Ltd., India
Pratik Katte	University of California, Santa Cruz, USA
Raghav Shrivastava	Picarro Inc., India
Ronak Dedhiya	Niramai Health Analytix Pvt Ltd., India
Sakhita Sree Gadde	Zurich North America, USA
Snekhalatha Umapathy	SRM Institute of Science & Technology, India

Contents

Video Transformers for Dynamic Breast Infrared Imaging Classification

Lenin G. Falconi[1]([✉])(iD), Maria Pérez[1](iD), Marco E. Benalcázar[1](iD), and Aura Conci[2](iD)

[1] Escuela Politécnica Nacional, Quito 170525, Ecuador
{lenin.falconi,maria.perez,marco.benalcazar}@epn.edu.ec
[2] Instituto de Computação, Universidade Federal Fluminense, Niterói, RJ 24210-346, Brazil
aconci@ic.uff.br

Abstract. Breast Cancer (BC) is a leading cause of mortality among women worldwide and early detection remains crucial for improving patient survival rates. Thermography offers a non-invasive imaging technique for identifying abnormal temperature patterns associated with malignancies. In this work, we propose a video transformer-based model for automatic BC classification from thermographic images using Universidade Federal Fluminense dynamic acquisition protocol. Our model achieved an Area Under the Receiver Operating Characteristic Curve score of 0.94 on a test set of 15 patients outperforming the results of the state-of-the-art set by 3D Convolutional Neural Networks.

Keywords: Breast cancer detection · Thermography · Transformers · Deep learning · Medical image analysis

1 Introduction

Breast Cancer (BC) remains one of the main health challenges affecting women worldwide and it is the second most lethal form of cancer in Ecuador [4]. Since its introduction in 1960, Mammography (MG) has remained the gold standard for early detection [13,25], with screening programs demonstrating effectiveness in reducing mortality rates. However, MG faces limitations: reduced efficacy in younger women with dense breast tissue, ionizing radiation risks, and reliance on manual interpretation of low-contrast lesions with varied morphologies [18,23]. In contrast, Thermography (TG) offers a radiation-free, non-invasive, and cost-effective alternative for BC screening that is applicable to all age groups without contraindications. Moreover, when combined with machine learning techniques, TG demonstrates significant potential for early BC detection [19].

Recent advances in Artificial Intelligence (AI), particularly in Deep Learning (DL), have significantly transformed computer-aided detection (CAD) systems for medical imaging [17]. Although convolutional neural networks (CNNs) currently dominate image processing, emerging transformer-based models demonstrate superior capabilities in global context modeling [8]. This study investigates

S. T. Kakileti et al. (Eds.): AIIIMA 2025, LNCS 16308, pp. 1–19, 2026.
https://doi.org/10.1007/978-3-032-10990-3_1

the application of transformer architectures, specifically Vision Transformers (ViTs), for automated classification of BC using the UFF's Dynamic Infrared Thermography (DIT) protocol, which employs a sequence of temperature maps per patient [27]. This Infrared (IR) image acquisition method established the superiority of dynamic protocols for highlighting vascular behavior [22]. In this work we hypothesize that ViTs can better model temporal dependencies in thermal sequences compared to CNNs. As a consequence, we examine both video and image-based ViTs implementations and compare their performance with that of CNNs-based architectures. Our primary objective is to improve diagnostic accuracy in BC classification through TG imaging. . The following are the proposed research questions for this research:

1. **RQ1:** Do ViTs outperform CNNs in BC classification from thermographic images due to their superior ability to model global contextual relationships?
2. **RQ2:** Does transfer learning with pre-trained ViTs architectures improve classification accuracy compared to models trained from scratch for this medical imaging task?
3. **RQ3:** Does combining pre-trained ViTs models with dynamic acquisition protocols enhance diagnostic classification performance for breast cancer detection in thermogram images compared to conventional approaches?

In this work, we fine tune a Video Vision Transformer (ViViT) to classify thermal images under the DIT protocol. Our study makes the following key contributions: (1) We extend Vision Transformer (ViT)-based approaches to DIT protocol, leveraging sequential image analysis to improve BC classification performance using a patient level dataset. (2) We evaluate the impact of Fine-Tuning (FT) using pre-trained ViTs. (3) We conduct experiments on two different versions of the DMR-IR dataset (thermal map and thermal image) and compare performance to a simple CNN architecture. (4) To the best of our knowledge, this is the first study accounting the use of Video Vision Transformers (ViViTs) in DIT protocol for TG in BC classification. Our model achieved an Area Under the Receiver Operating Characteristic Curve (AUC)-score of 0.9350 ± 0.0094 in 10 fold cross validation setup and 0.9444 on a separated test set of 15 patients with stratified labels. This work advances the state-of-the-art set by [6], where 3D CNNs achieved 93.51% AUC in binary classification, and compliments the work by [11], which achieved 95.7% in Static Infrared Thermography (SIT) protocol [22, 27].

The organization of this paper is presented as follows: In Sect. 2, a brief introduction to main characteristics about the TG protocols for screening BC is presented. Section 4 presents the related works to our research questions. Our proposed methodology is presented in Sect. 5. Results are presented in Sect. 6. Finally, discussion over main findings is presented in Sect. 7 and this work's conclusions in Sect. 8.

2   Advantages of Dynamic Infrared Thermography in Breast Cancer

By acquiring the body's natural emissions of infrared radiation to map the body's temperature distribution, TG is a non-invasive imaging modality that facilitates BC detection [13,19,25]. Due to the heightened metabolic activity of cancerous cells, there is an increased blood flow to the tumor, which promotes the development of new blood vessels presenting abnormal behavior under temperature modifications. These anomalies typically appear as thermal asymmetries between the contralateral sides in the breast region under induced thermal stress.

Table 1. Comparison of MG and TG

Parameter	Mammography	Thermography
Basis	X-rays	Infrared radiation
Device	X-ray machine[3]	Infrared camera
Output	Anatomical image	Thermal map
Abnormality Type	Anatomical	Physiological
Equipment cost	$\approx$ \$80K[a]	$\approx$ \$0.4K[b]
Age	over 40 years	No age barrier
Women with implants	Not so effective	Effective
Breast Density	False Positives	Not affected
Further Investigation	Unnecessary biopsies	Fewer biopsies
Pain and Fear	Breast Compression	No touch
Radiation	Ionizing radiation	None
FDA approval	Cancer diagnosis	Adjunct to MG

[a]https://www.blockimaging.com/
[b]https://www.flir.com.mx

Moreover, unlike MG, which is less effective in younger patients due to the characteristics of denser breast tissue, TG can be safely applied across all age groups and is generally more cost-effective than MG and other imaging modalities. It can also be applied to females with dense breast tissue, women who are pregnant or breast feeding. Although the FDA approved TG as a cancer risk assessment tool in 1982, it is typically used as an adjunct to MG rather than a standalone diagnostic method [13,20]. Advances in thermal sensing equipment, image processing, machine learning and AI pipelines have contributed to more feasible and accurate CAD systems for infrared TG.

TG images can be obtained by using two different protocols: SIT and DIT. The former consists in capturing the thermal images under controlled conditions of frontal, oblique and lateral breast views [20]. The latter measures the thermal response to a thermal stress applied to the patient [20,25]. The UFF's DIT protocol captures an image every 15 s over an interval of 5 min, resulting in a sequence of 20 images per patient [6,27]. Despite its advantages, TG is susceptible to procedural inconsistencies and thermal noise interference. A major challenge lies in the non-specificity of thermal anomalies, which complicates accurate interpretation. For instance, elevated temperatures may result from non-cancerous conditions such as infections or inflammations. A comparative overview of TG and MG is presented in Table 1.

3   Approaches for Sequence Modeling

3.1   Image Vision Transformers

In 2017, the Transformer architecture was introduced in the seminal paper "Attention Is All You Need" as the first sequence transduction model relying entirely on attention mechanisms [26]. This architecture replaced recurrent layers with an encoder-decoder structure based on multi-head self-attention, establishing itself as the state-of-the-art approach for Natural Language Processing (NLP) tasks.

Attention mechanisms had been previously explored in image processing [8]. However, this work demonstrated that a pure Transformer architecture could be directly applied to sequences of image patches, treating them analogously to tokens in NLP tasks with minimal modifications to the original architecture. An image patch, in a ViT is a fundamental hyperparameter that determines how an input image is divided into smaller segments (patches) before being processed by the transformer architecture. The patch size defines the dimensions of each non-overlapping block that the image is split into. For an input image of size $H \times W \times C$ (Height $\times$ Width $\times$ Channels) and a patch size of $P \times P$, the number of patches is $N = \frac{H}{P} \times \frac{W}{P}$; where each patch becomes a token of dimension $P \times P \times C$. For instance, if the image is $(224 \times 224 \times 3)$, $N = \frac{224}{16} \times \frac{224}{16} = 14 \times 14 = 196$ tokens.

Dosovitskiy et al. showed that ViTs achieve superior performance when pretrained on large-scale datasets such as ImageNet-21K and JFT-300M, followed by FT for downstream tasks, outperforming state-of-the-art CNNs. Additionally, their work provides key implementation insights: (1) **Positional embeddings** show no significant improvement when extended to 2D representations; (2) **FT** employs higher-resolution images while maintaining original patch size [15]; (3) **Transfer Learning (TL)** replaces the classification head with a zero-initialized linear layer matching the target output dimension; (4) **Training** requires strong regularization for scratch training on ImageNet with cosine learning rate decay; (5) **Preprocessing** adheres to guidelines from [15].

3.2   Video Transformers

Video can be understood as a sequence of images through a time dimension that introduces motion and deformations on them [21]. Because of this sequential nature present in video information, it would be expected that ViTs to be effective in video modeling [3]. The *TimeSformer* stands as the first convolution-free architecture that adapts the ViT architecture proposed in [8] to video by extending the self-attention mechanism from the image space to the space-time 3D volum [3]. Their main idea is to treat video as a sequence of patches extracted from the individual frames, where each patch is linearly mapped into an embedding space with positional information [3].

There is active research on ViViTs, with several notable models proposed in recent literature (e.g., ViViT [1], MViT [10]). Most improvements across different architectures primarily address the computational complexity inherent in Transformer models, which scales quadratically with respect to the sequence length T (i.e., $\mathcal{O}(T^2)$). Additionally, these approaches tackle challenges related to spatial redundancy, temporal fidelity, and inductive biases [21].

3.3 CNN-Based Sequence Modeling

While Recurrent Neural Networks (RNN) architectures (e.g., LSTM and GRU) have traditionally been the default choice for sequence modeling, recent studies suggest that CNNs may outperform them in certain tasks [2]. The Temporal Convolutional Network (TCN) is a general-purpose convolutional architecture for sequence prediction characterized with two key properties: (1) *causal convolutions*, which prevent information leakage from future to past data, and (2) a *fixed output length*, enabling the model to map variable-length inputs to outputs of the same length [2].

A significant advantage of TCN is its inherent parallelism, where predictions for later time steps do not depend sequentially on earlier ones. This is achieved by processing the same filter across each layer, allowing the model to handle long input sequences efficiently during both training and evaluation.

In contrast, Keras [5], a popular DL framework, provides a Time Distributed Layers (TDL) wrapper that applies a given layer to every temporal slice of an input. By reusing the same weights across all timestamps in the 3D input tensor, this approach ensures parameter efficiency. Recent work by [7] demonstrates that combining TDL with LSTM can simultaneously improve feature extraction and temporal pattern recognition in sequential data, leading to enhanced accuracy in (HAR) tasks.

4 Literature Review

This section provides a systematic literature review following the methodology proposed in [14] to find relevant works in the area of this research goals and questions. To ensure comprehensive coverage of ViTs and TG applications, we employed the Population, Intervention, Comparison, Outcome, Context (PICOC) framework to engineer the search string. We focused specifically on studies that combined thermographic imaging with advanced computer vision techniques, particularly ViTs, in the context of BC detection.

4.1 Search Methodology

We conducted targeted searches across four major digital libraries, such as: ACM Digital Library, IEEE Xplore, Scopus, and Springer Link. Additional repositories were examined but yielded no relevant results aligned with our research scope. We employed the following Boolean search query:

```
(thermogram OR thermography) AND
(vision transformer OR vit) AND
(cnn OR convolutional neural networks OR deep learning OR machine learning) AND
(improve OR enhance) AND
(classification OR detection OR diagnosis OR segmentation) AND
breast cancer
```

The search query was designed to identify research papers comparing ViTs and CNNs models. Given the scope of our study, we adopted broad inclusion criteria for DL-based computer vision models applicable to TG image classification. We systematically excluded studies that did not primarily focus on TG as the screening modality, such as those employing ultrasound, magnetic resonance imaging, or MG. Furthermore, non-specific sources (e.g., journals containing multiple unrelated articles) were discarded. During the search for BC research, we also encountered numerous studies on histopathological and genetic analyses, which were outside our scope and thus excluded.

The execution of the proposed search strategy yielded 24 primary studies after removing duplicates. Subsequently, each study underwent a screening process where titles, keywords, and abstracts were evaluated against the predefined inclusion and exclusion criteria. This selection process resulted in 4 relevant publications for the literature review. The scarcity of publications investigating ViT and TG suggests significant opportunities for novel contributions, with only one study [11] addressing research objectives comparable to ours.

4.2 Deep Learning for Dynamic Breast Thermography

According to our proposed search strategy for literature review, the first study employing ViTs for BC classification using TG images is reported in [11]. Their approach utilizes the standard ViT-Base architecture with 12 layers, 12 attention heads, and a 768-dimensional embedding spac [8]. Similar to our work, their methodology leverages the transformer's self-attention mechanism for TG image classification into *Cancerous* or *Healthy* categories. The authors investigate the impact of patch size on classification accuracy, demonstrating that 32×32 patches yield superior performance compared to the standard 16×16 configuration. Also, it is worth to mention that their implementation does not incorporate pre-trained weights on natural images or data augmentation techniques. Their approach achieved an AUC of 95.7% for binary classification on the DMR-IR datase [22], using a total of 950 samples with an analysis unit per TG image. However, the study was limited to SIT acquisition protocol.

To benchmark our methodology against CNNs-based approaches for DIT, we compare with [6], which employs 3D-CNNs to recognize patterns changes in a sequence of TG images. Cancer identification using a sequence of images is similar to action recognition and depend on spatio-temporal models that extract space-time feature information from data. While 2D CNNs applied to individual frames remain competitive for action recognition [24], 3D convolutions inherently capture motion information across adjacent frames [12,16]. In [6], each patient's data is represented as a hypercube constructed by stacking 20 preprocessed frames. The preprocessing stage involved the segmentation of the region of interest from the original TG images and their resized to $32 \times 32 \times 3$. Baffa et al. enhanced the model generalization through data augmentation with the following operations: rotations up to $15°$, horizontal flips, shearing transformations and adding noise. According to them, 3D CNN is capable of considering the spatial relationships in each image and their evolution over a temporal dimension. Their experiments employed the DMR-IR dataset, comprising $2\,980$ images from 149 patients. Under a K-fold cross-validation strategy, their model achieved an AUC score of $93.51\% \pm 4.03\%$.

Despite the current trend of researching modern DL architectures for BC classification in infrared TG, it should be noted that texture-based CAD systems can accurately classify BC in TG, achieving over 90% accuracy, due to the rich texture content present in the images [20]. A significant advantage of these texture-based methods is their computational efficiency, particularly when dealing with limited datasets, as they require substantially fewer resources than their DL counterparts. This efficiency, however, comes at the cost of requiring manual feature engineering, whereas DL approaches automate the feature learning process.

5   Methodology

This section outlines the methodological procedures employed to develop and validate our set of models for BC using the DIT protocol, specially the efficacy of ViTs models. The most similar work to our approach was proposed in [11], where the authors trained a ViT model from scratch using the same dataset but the SIT protocol. Their study does not evaluate the performance of pre-trained ViTs, which is studied now here.

From our literature review there is no previous automated classification of BC by using DIT protocol and ViViTs. To benchmark our results, we consider the work of Baffa et al. [6], where 3D CNNs are employed to study the evolution of the sequence of TG images.

The methodology consists of seven main stages (Fig. 1), where some steps vary depending on the specific experiment. First, we formulated the machine learning problem. Thereafter, we retrieved the sequences of Temperature Maps (TM) ($\mathbb{D}_{TM} \in \mathbb{R}^{640 \times 480 \times 1}$) and IR images ($\mathbb{D}_{IR} \in \mathbb{Z}^{640 \times 480 \times 3}$) for each subject from the DMR-IR dataset, creating two distinct data subsets. The TM dataset comprises temperature matrices with a precision of up to two decimal places and an accuracy of $0.02°C$, whereas the IR dataset consists of three-channel RGB infrared images with pixel values ranging from 0 to 255.

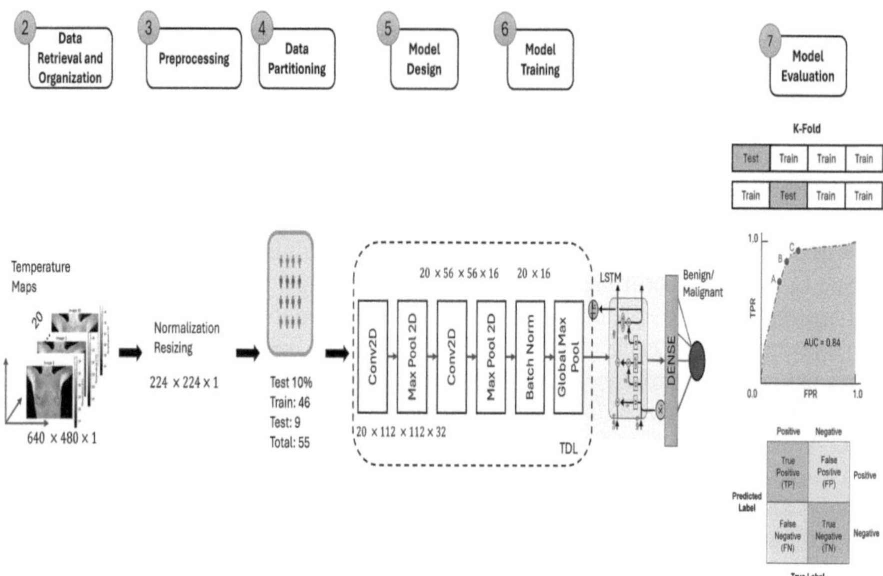

(a) Methodology for Temperature Maps: depicts the CNN network with TDL

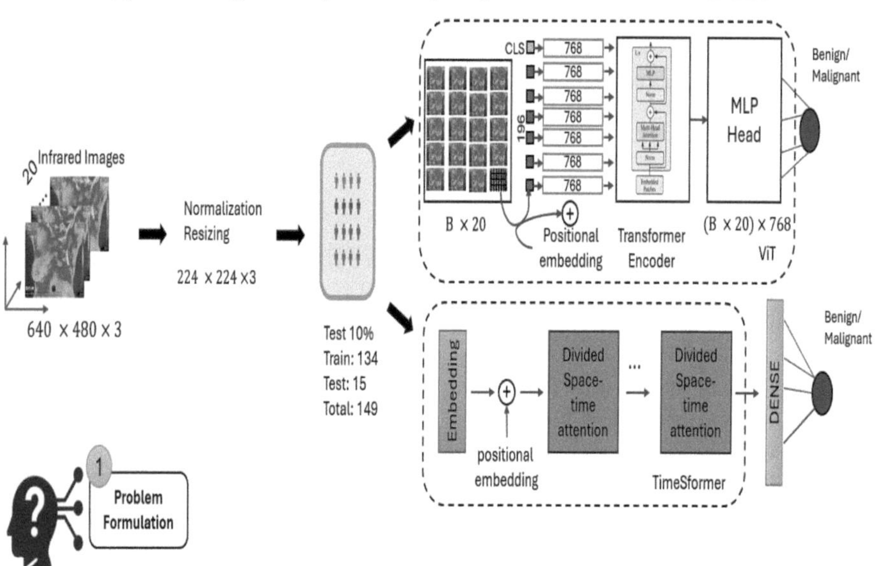

(b) Methodology for Infrared Thermography Images: depicts the ViT and ViViT models

Fig. 1. Overview of the methodology used in this research. It comprises seven stages: (1) Problem Formulation, (2) Data Organization, (3) Preprocessing, (4) Data Partitioning, (5) Model Design, (6) Model Training, and (7) Model Evaluation for (a) Temperature Maps and (b) Infrared Thermography Images.

Next, we set the preprocessing pipelines for each dataset according to its specific characteristics, including data augmentation during training. Afterward, for $\mathbb{D}_{\mathrm{IR}}$, we applied a stratified holdout sampling technique, reserving 10% of the cases as a test set. We then designed the models using a 5-dimensional input tensor $\mathcal{X} \in \mathbb{R}^{B \times F \times C \times H \times W}$, where: B denotes the batch size, F represents the number of frames, C corresponds to the number of channels, and H and W indicate the height and width of the TM or IR images, respectively. Due to the small dataset size, we evaluated the models using K-fold cross-validation. Finally, we selected the best-performing model and evaluated it on the test set. We employed AdamW as the optimizer together with checkpoint and early stopping callbacks. All experiments were performed using Kaggle's GPU 1 x Tesla P100 of 3584 CUDA cores, 16 GB GDDR6 and up to 30 GB of RAM.

5.1   Problem Formulation

Our objective is to train a model to classify a sequence of TG data into either *Benign* or *Cancer*. Consequently, we want to find a mapping function $f_\theta : \mathbb{R}^{F \times C \times H \times W} \to [0, 1]$ with parameters θ, where the predicted class probabilities for an input sequence $\mathbf{x} \in \mathcal{X} \subset \mathbb{R}^{F \times C \times H \times W}$:

$$P(y = 1 \mid \mathbf{x}) = f_\theta(\mathbf{x}) \quad (\text{for } \mathbf{x} \in \mathcal{X}) \tag{1}$$

Considering that a video classification aims to predict the class of a given input sequence of frames [21], the final prediction $\hat{y} \in \{0, 1\}$ is:

$$\hat{y} = \mathbb{I}\left(P(y = 1 \mid \mathbf{x}) \geq 0.5\right) \tag{2}$$

The optimization minimizes the binary cross-entropy loss:

$$\mathcal{L}(\theta) = -\frac{1}{N} \sum_{i=1}^{N} \left[y_i \log(\hat{y}_i) + (1 - y_i) \log(1 - \hat{y}_i) \right] \tag{3}$$

where $y \in \{0, 1\}$ are the ground truth labels.

In this context, we hypothesize that ViViTs can outperform 3D CNNs in BC classification for TG.

5.2   Model Fine-Tuning

FT is a TL technique where the weights of a pre-trained model are further trained on a smaller, task-specific dataset. Typically, the classification head is replaced to adapt the model to the nuances of the new task. This process can update all layers of the pre-trained model or gradually unfreeze some of them (layer-wise fine-tuning). We adopt the formal definitions for TL and FT given in [9]:

Definition 1 (Transfer Learning). *Given an original domain $\mathcal{D}^S$ with an original learning task $\mathcal{T}^S$, a target domain $\mathcal{D}^T$ with a target task $\mathcal{T}^T$, the transfer* *learning operation seeks to improve the learning of a prediction function $\phi^{\mathrm{T}}(\cdot)$ in $\mathcal{D}^T$ using the knowledge acquired in $\phi^{\mathrm{S}}(\cdot)$ in $\mathcal{T}^S$, through ϕ^S; where $\mathcal{D}^S \neq \mathcal{D}^T$ or $\mathcal{T}^S \neq \mathcal{T}^T$.*

Equations (4) and (5) represent the transfer learning operation and model prediction, respectively. Here, L denotes the number of layers in the pre-trained model, and $\mathcal{A}$ refers to a set of operations that replace the original classification head.

$$\mathbb{T}_L \left\langle \phi^{\mathrm{S}}(\mathcal{D}^S), \mathcal{D}^T \right\rangle = \mathcal{A}\left(\phi^{\mathrm{S}}(\mathcal{D}^T)\Big|_0^L \right) \tag{4}$$

$$\mathcal{Y}^T = \mathbb{T}_L \left\langle \phi^{\mathrm{S}}(\mathcal{D}^S), \mathcal{D}^T \right\rangle = \phi^{\mathrm{T}}(\mathcal{D}^T) \tag{5}$$

Definition 2 (Fine-Tuning). *Given an original domain $\mathcal{D}^S$ with an original learning task $\mathcal{T}^S$, a target domain $\mathcal{D}^T$ with a target task $\mathcal{T}^T$, the fine-tuning procedure enhances the learning of a target function $\mathcal{Y}^T = \phi^{\mathrm{T}}(\mathcal{I}^T)$ by retraining r layers of the original model $\mathcal{Y}^S = \phi^{\mathrm{S}}(\mathcal{I}^S)$ with L layers, where $\gamma = L - r$.*

Equations (6) and (7) represent the fine-tuning operation and the fine-tuned model prediction, respectively. Full model update occurs when $r = L$. The operation $\mathcal{A}$ involves replacing the original classification head of the pre-trained model. For instance, in the case of TimeSformer, it would replace the original 400-class output for Kinetics-400 with a new head appropriate for the target task.

$$\mathbb{F}_T \left\langle \phi^{\mathrm{S}}(\mathcal{D}^S), \mathcal{D}^T \right\rangle = \mathcal{A}\left(\phi^{\mathrm{S}} \left(\phi^{\mathrm{S}}(\mathcal{D}^T)\Big|_0^{\gamma-1} \right) \Big|_\gamma^L \right) \tag{6}$$

$$\mathbb{F}_T \left\langle \phi^{\mathrm{S}}(\mathcal{D}^S), \mathcal{D}^T \right\rangle = \mathcal{A}\left(\phi^{\mathrm{S}}(\mathcal{D}^T) \right) \tag{7}$$

5.3   Data Collection, Preprocessing and Data Augmentation

DMR-IR dataset contains TG data of BC patients acquired under either the SIT or the DIT protocol. This data is provided in two formats: TM and IR images. The former comprises a singleâĂŞchannel, twoâĂŞdimensional matrix of floating-point temperature values captured by the FLIR SC620 infrared camera with a measurement accuracy of $\pm 2\,^\circ$C [6]. The latter consists of RGB images acquired from the same camera. In both cases, the resolution is 640×480 pixels. However, as noted in [11], the number of subjects and images varies across studies, making direct comparisons difficult. Table 2 summarizes the characteristics of $\mathbb{D}_{\mathrm{TM}}$ and $\mathbb{D}^{\mathrm{IR}}$.

Even though all images in $\mathbb{D}_{\mathrm{IR}}$, are .jpg files, we found that 28 cases presented images of 160×120 and 1 case of 800×600. All of which were resized to 640×480. Also, a total of 21 images were copies of a previous sequence. This happened to 7 cases.

For $\mathbb{D}_{\mathrm{TM}}$, normalization was performed using the entire patient sequence. As for the case of $\mathbb{D}_{\mathrm{IR}}$, predefined normalization values were applied to match

Table 2. DMR-IR dataset characteristics

Feature	$\mathbb{D}_{TM}$	$\mathbb{D}_{IR}$
Type	Temperature Map	Infrared RGB Image
Matrix size	640×480	640×480
Channels	1	3
Number type	float(temperature)	integer(pixel)
Total patients	55	149
- With cancer	36	87
- Without cancer	19	62
Frames per patient	20	20
Total images	1 100	2 980
Training set patients	46	134
Test set patients	9	15

the requirements of the pre-trained networks employed. Since the DIT protocol operates at the patient level, the number of available samples is inherently limited. To mitigate this constraint, data augmentation was applied exclusively to the training fold. The augmentation strategy differs from that of [6], incorporating only the following transformations: horizontal flip, rotation up to $15°$, and resize. Other augmentation techniques were excluded to preserve the thermal properties of the data, which differ significantly from natural images. Building upon this idea, we do not consider of use to apply shearing transformations and introducing noise.

5.4   Model Development

Two categories of models were studied: (1) spatio-temporal architectures, and (2) spatial-only. The former experiments with TDL and ViViTs. While, the latter, applies the basic ViT designed to the flatten inputs (i.e. $(B \times F, C, H)$) without temporal modeling. Both PyTorch and Keras/Tensorflow were used as the deep learning backends. Table 3 presents the models studied, along with the source domain dataset used when FT was applied for model adaptation. The input tensor dimensions, deep learning backend, and number of parameters are also provided.

Time Distributed Layer Model: Building upon the hybrid architecture proposed in [7], we implement a sequential model combining TDL for spatial feature extraction and Long Short Memory (LSTM) for temporal modeling. Each frame undergoes spatial feature extraction through resizing to 224×224 resolution, processing through two convolutional blocks (32 and 16 filters with 3×3 kernels and ReLU activation), max pooling (2×2) after each convolution, and batch normalization for stable training. Subsequently, temporal processing is performed via global max pooling to reduce spatial dimensions to 16 features per frame, followed by an LSTM layer (16 units) with ℓ_2 regularization ($\lambda = 0.01$) to model temporal dynamics. Finally, the classification head consists of a dense layer (16 units, ReLU) with ℓ_2 regularization, dropout ($p = 0.5$) for additional regularization, and a sigmoid output for binary classification.

The TDL wrapper employs weight sharing across all frames, applying identical transformations to each temporal slice. This architecture efficiently integrates spatial feature learning through convolutional operations, temporal modeling via recurrent connections, and regularization techniques (ℓ_2 and dropout) to mitigate overfitting. During training, data augmentation was applied, and early stopping with model checkpointing was employed based on validation loss. This model was engineered to account for the lack of patients in the $\mathbb{D}_{TM}$ dataset. It has a total of 7 377 trainable parameters and was coded using Keras/Tensorflow. Resizing of TM was carried out due to memory constrains.

Basic Vision Transformer: ViTs are inherently spatial architectures. We designed our ViT-based classifier to process complete sequences of 20 thermal frames per case, maintaining compatibility with the standard 5D input tensor format (B, F, C, H, W). We construct our model from the base architecture proposed in [8].

The architecture comprises three principal components: (1) a pre-trained ViT-B/16 backbone, from which the classification head has been removed; (2) frame-wise feature extraction where input sequences are reshaped from (B, F, C, H, W) to $(B \times F, C, H, W)$, processing each frame independently through the ViT encoder and extracting features vectors from the final layer with $D = 768$; (3) a linear classification head that projects these features to logits of shape $(B \times F, 1)$, and computes temporal mean predictions $(B, 1)$.

This design enables parallel processing of all frames while preserving sequence information through temporal averaging. The model accepts the same 5D tensor format as traditional video architectures but leverages ViT's superior spatial modeling capabilities for thermal image analysis. A total of 85 799 425 parameters are to be trained, when using 16×16 patches and 87 456 001, when the size of the patches is increased to 32×32. We experimented the ViT with initialized ImageNet1K weights and from scratch for the latter case.

Video Transformer. Since video classification requires both spatial and temporal processing of image sequences, we implement a TimeSformer-based architecture [3] for DITanalysis. Generally, models for video data need to address two key challenges: (1) joint spatio-temporal feature learning, and (2) handling variable-length sequences. However, in DIT, this second challenge is mitigated by the standardized protocol where each patient examination contains exactly 20 IR frames.

Our   TimeSformerThermoClassifier model comprises a pre-trained TimeSformer-base model, which was finetuned on Kinetics-400, and a modified classification head. The TimeSformer model is configured to process an input of $F = 8$ frames at $H \times W$ resolution; it employs spatio-temporal attention across divided space-time patches and maintains a hidden dimension of $D = 768$ across all layers. For the classification head, the original Kinetics-400 classification layer is replaced with a single-output linear layer to perform binary prediction; a sigmoid activation is implicitly used through the binary cross-entropy loss.

The forward pass processes input tensors $\mathbf{X} \in \mathbb{R}^{B \times F \times C \times H \times W}$ through a sequence of operations: first, frame rearrangement is applied when needed to match TimeSformer's expected (B, C, F, H, W) format; subsequently, spatio-temporal feature extraction is performed via patch embedding (typically 16×16) and divided space-time attention layers; following this, global average pooling of the spatio-temporal features is conducted; and finally, binary classification is achieved through the final linear layer.

This architecture leverages TimeSformer's proven effectiveness in video understanding while adapting it specifically for thermal medical imaging, where temporal consistency is guaranteed by the DIT acquisition protocol. However, the base architecture used in this research is able to hold up to 8 frames. To account to this fact we developed a random sampling from sequence code which allowed to extract up to 8 frames from the 20 without altering the order of the sequence.

Table 3. Specifications of the source domains (pre-training datasets), target domain (fine-tuning dataset), input tensor, framework, and parameters for each model variant tested.

Model	Pre-Trained	Source Domain $\mathcal{D}^S$	Target Domain $\mathcal{D}^T$	Input Tensor	Framework	# Parameters
ViT(16×16)	Yes	ImageNet-1K	$\mathbb{D}_{\text{IR}}$	$(B \times 20, 3, 224, 224)$	PyTorch	85 799 425
ViT(32×32)	Yes	ImageNet-1K	$\mathbb{D}_{\text{IR}}$	$(B \times 20, 3, 224, 224)$	PyTorch	87 456 001
ViT(32×32)	No	None	$\mathbb{D}_{\text{IR}}$	$(B \times 20, 3, 224, 224)$	PyTorch	87 456 001
ViViT / TimeSformer	Yes	Kinetics-400	$\mathbb{D}_{\text{IR}}$	$(B, 8, 3, 224, 224)$	PyTorch	121 259 521
TDL + LSTM	No	None	$\mathbb{D}_{\text{TM}}$	$(B, 20, 1, 224, 224)$	TensorFlow	7 377

5.5   Experimental Protocol

Figure 2 presents a guide for the experiments carried out in this research and compares the performance of the different models on the *F1-score* metric for the *K-Fold Cross Validation* strategy. Sequential models are highlighted.

Our experimental protocol varied across modalities and architectures. For TM data, we employed 3-fold cross-validation trained for 50 epochs using AdamW optimizer while monitoring validation loss. For IR images with pre-trained ViT models (both 16×16 and 32×32 patch sizes), we used 10-fold cross-validation for 20 epochs with a learning rate (lr) of 3e$-$5, optimizing for validation F1-score. However, when training the 32×32 patch ViT from scratch, we extended training to 50 epochs with higher initial lr (1e$-$4), weight decay (0.05),

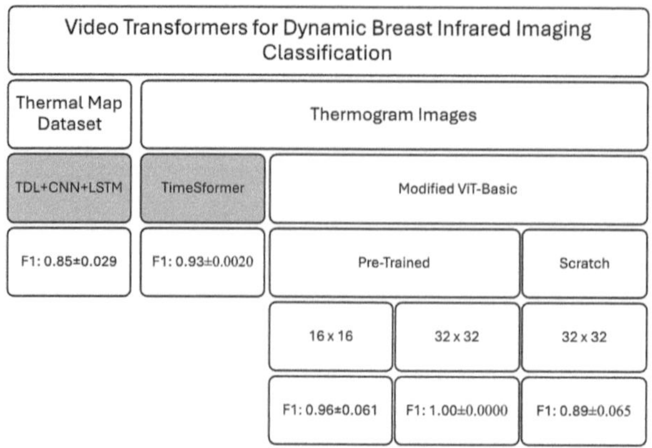

Fig. 2. A comparison of F1-score performance of all the Models for BC in DIT in this research. Highlighted are the sequential models like TDL and ViViT

Table 4. Training protocols for different modalities and architectures

Modality	Model	Training Configuration	Validation	Epochs
TM	TDL+CNN+LSTM	AdamW, $k = 3$ CV	Loss	50
IR	ViT (16×16) pre-trained	AdamW, lr $=$ 3e−5, $k = 10$ CV	F1-score	20
	ViT (32×32) pre-trained	AdamW, lr $=$ 3e−5, $k = 10$ CV	F1-score	20
IR	ViT (32×32) scratch	AdamW, lr $=$ 1e−4, wd $= 0.05$ LR warmup $+$ cosine (min lr $=$ 1e−6), $k = 10$ CV	Loss	50
IR	ViViT	AdamW, lr $=$ 3e−5, $k = 10$ CV	F1-score	20

and implemented lr warmup with cosine annealing (minimum lr $=$ 1e−6), while monitoring validation loss. The ViViT model followed similar configuration to pre-trained ViTs (20 epochs, lr $=$ 3e−5) monitored by F1-score. Table 4 presents the training protocols used for the different modality-architecture combinations.

6   Results

In this section, we discuss the quantitative performance of the different models in our proposed experiments (Table 5) and compare our work with the most relevant studies identified through our literature review (Table 6).

Our work is similar to that of Garia et al. [11] in the use of Transformer-based models for breast cancer classification using Thermography (TG). However, we address dynamic infrared thermography (DIT), which requires a different methodological focus. Specifically, we investigate: (1) a temporal deep learning (TDL) approach using Time Distributed Layer (TDL)+CNNs+LSTM; (2) vision transformers (ViT), both with pre-trained weights and trained from scratch; and (3) pre-trained video vision transformers (ViViT).

Consequently, our problem is most similar to the work of Baffa et al. [6]. Unlike their approach, we propose the use of pre-trained ViViT TimeSformer to address the sequential nature of DIT. Our results indicate that ViViTs exhibit performance comparable to 3D-CNNs for UFF's dynamic acquisition protocol in the DMR-IR dataset [27]. However, there are important differences with respect to their work: (a) we work with a larger input tensor, providing higher resolution (their work resizes inputs to $(B, 20, 32, 32, 3)$, while we use $(B, 8, 224, 224, 3)$); (b) although the TimeSformer architecture reduces the number of frames processed, the model achieves similar performance.

Similar to Baffa et al., our unit of analysis is the patient, whereas Garia et al. focus on individual images since they focus on SIT protocol [6,11]. Experiments using the basic ViT architecture show performance comparable to ViViT, specially when using 32×32 patches. But it is important to notice that ViT cannot process infrared (IR) images sequentially because they are inherently spatial and do not operate across the time dimension. As a result, frames are effectively averaged during processing.

Finally, our results suggest that further investigation is required when working directly with temperature matrices (TM). We also recommend that future research clearly indicate whether they are using IR images or raw temperature values.

7   Discussion

To the best of our knowledge, this work presents the first application of a ViViT to the DIT protocol. Our approach achieves superior performance on a test set of 15 patients (93% accuracy, 93% precision, 94% recall) compared to the results reported in [6]. Furthermore, in a $K = 10$ cross-validation setting, our method achieves a similar mean accuracy of 93.50% but demonstrates superior stability, as indicated by a significantly smaller standard deviation of $\pm 0.94\%$. This higher precision and lower variance suggest a more robust model. It is noteworthy that our experiments utilize IR images with a spatial resolution of 224×224 pixels, a significantly larger input size compared to the 32×32 images used by the baseline method in [6]. We employ a shorter temporal sequence ($F = 8$ frames) to accommodate the use of a pre-trained model. To address this, future ablations studies should quantify the incremental value of temporal modeling by varying (a) the number of frames (e.g., 3, 4, 6, 8, and the full 20-frame sequence), (b) the sampling strategy (contiguous vs. uniformly spaced frames), and (c) the timing (early vs. late frames relative to cooling).

Table 5. Performance comparison of different models on DIT data

Dataset	Model	Metric	Cross-Val (Mean ± Std)	Test Set
$\mathbb{D}_{\text{TM}}$	TDL+CNN+LSTM	F1-score	0.8585 ± 0.0296 ($k = 3$)	-
		AUC	-	-
$\mathbb{D}_{\text{IR}}$	ViT (16×16) pre-trained	F1-score	0.9644 ± 0.0458	-
		Accuracy	0.9538 ± 0.0615	0.87
		AUC	-	-
	ViT (32×32) pre-trained	F1-score	1.0000 ± 0.0000	0.9412
		Accuracy	1.0000 ± 0.0000	0.9333
		AUC	1.0000 ± 0.0000	0.9444
	ViT (32×32) scratch	F1-score	0.9101 ± 0.0665	0.8889
		Accuracy	0.8939 ± 0.0652	0.8667
		AUC	0.8866 ± 0.0595	0.8611
	ViViT	F1-score	0.9339 ± 0.0020	0.9412
		Accuracy	0.9231 ± 0.0000	0.9333
		AUC	0.9350 ± 0.0094	0.9444

Table 6. Comparison of studies on breast thermography classification

Study	Model	Protocol	Pre-trained	Dataset	Analysis Unit	Total Subjects	Cross Validation	Test Set	AUC
Garia et al. [11]	ViT	SIT	No	DMR-IR	Images	950	-	10%	0.9570
Baffa et al. [6]	3D-CNN	DIT	No	DMR-IR	Patient	149	K=10	-	0.9351
Our work	ViViT	DIT	Yes	DMR-IR	Patient	149	K=10	10%	**0.9444**

Another important difference is that our current results do not use a prior segmentation stage. Moreover, this arises the question if segmentation would increase performance in our results. However, as indicated in [20], there are no ablation studies that could confirm this in infrared TG.

Our experiments corroborate that ViTs with 32×32 patches yield better performance. However, it should be noted that a different acquisition protocol for the same dataset was employed compared to [11]. In contrast to their findings, pre-training clearly benefited performance in our experiments, despite the domain gap between natural images/videos and medical thermal data.

DIT aims to capture physiological responses under thermal stress that may not be visible in single static frames. Our results show that a pre-trained spatial ViT with temporal averaging can reach performance comparable to a pre-trained video transformer, suggesting that a substantial portion of the discriminative signal may be spatial. This raises two complementary questions: (i) whether the current DIT protocol sufficiently captures the temporal dynamics it is designed to measure, and (ii) whether simpler static protocols (SIT) augmented with robust spatial modeling could be adequate for certain clinical scenarios.

The results also suggest that future works should emphasize the type of thermal data used, as outcomes obtained directly from the point temperatures in CSV files differ from those using IR images. The challenges associated with temperature maps may stem from the fact that pre-trained networks are typically trained on natural images, which constitute a domain distinct from both IR imagery and temperature maps. We observed that infrared RGB images delivered stronger performance than raw temperature matrices.

7.1 Threats to Validity

Our study is primarily limited by: (i) the relatively small size of the dataset, and (ii) the absence of clinical validation. Although this research prevented patient information leakage in the hold-out test set, the limited sample size, combined with the use of large architectures such as transformers, increases the risk of overfitting. This risk, in turn, limits the potential for the clinical translation of these results. To mitigate overfitting, FT and data augmentation were employed; however, future work should include statistical tests and ablation studies to further validate and improve the model's performance. The sample size of available public datasets, however, is a factor over which we have limited control, besides some of the suggestions indicated in [20]. Clinical validation, while challenging to achieve, is necessary. It would not only validate the current model but also enhance it by incorporating other physiological factors (e.g., menstrual cycle, hormone therapy) that may influence thermographic patterns and, consequently, the model's performance.

8 Conclusions

This research paper shows that ViViTs can be effectively applied to BC classification using dynamically acquired TG images. Our current results show that the basic TimeSformer with 8 frames has achieved a similar performance to 3D-CNN in a prior work. Also, this work extends that of [11] by extending to the dynamic protocol.

With respect to the research questions, we conclude that transformer-based models exhibit a performance comparable to that of CNNs for this research problem. In our experiments, fine-tuning a pre-trained model yielded superior performance. Finally, the integration of DIT with ViViTs achieves results similar to those reported in [11]. Moreover, our results indicate that ViTs with temporal averaging perform similarly to fine-tuned ViViTs. This suggests that further research is necessary to fully account for the effect of temporal dynamics in BC prediction using infrared images and transformer-based architectures.

Acknowledgments. Maria Pérez, Marco E. Benalcázar and Lenin G. Falconi would like to thank the support of Escuela Politécnica Nacional for the funding of this research. Aura Conci thanks the Brazilian agencies (FAPERJ and CNPq) for funding this research at UFF (307638/2022-7 and E26/200.924/2022).

Disclosure of Interests. The authors have no competing interests to declare that are relevant to the content of this article

References

1. Arnab, A., Dehghani, M., Heigold, G., Sun, C., Lučić, M., Schmid, C.: Vivit: a video vision transformer. In: Proceedings of the IEEE International Conference on Computer Vision, pp. 6816–6826 (2021). https://doi.org/10.1109/ICCV48922.2021.00676, https://arxiv.org/pdf/2103.15691
2. Bai, S., Kolter, J.Z., Koltun, V.: An empirical evaluation of generic convolutional and recurrent networks for sequence modeling (2018). https://arxiv.org/pdf/1803.01271
3. Bertasius, G., Wang, H., Torresani, L.: Is space-time attention all you need for video understanding? Proc. Mach. Learn. Res. **139**, 813–824 (2021). https://arxiv.org/pdf/2102.05095
4. Bray, F., Ferlay, J., Soerjomataram, I., Siegel, R.L., Torre, L.A., Jemal, A.: Global cancer statistics 2018: GLOBOCAN estimates of incidence and mortality worldwide for 36 cancers in 185 countries. CA: A Cancer J. Clinicians **68**(6), 394–424 (2018). https://doi.org/10.3322/caac.21492, http://doi.wiley.com/10.3322/caac.21492
5. Chollet, F., et al.: Keras (2015). https://keras.io
6. de Freitas Oliveira Baffa, M., Lattari, L.G., Conci, A.: 3D convolutional neural networks for dynamic breast infrared imaging classification. In: Kakileti, S.T., Manjunath, G., Schwartz, R.G., Frangi, A.F. (eds.) Artificial Intelligence over Infrared Images for Medical Applications, pp. 57–66. Springer Nature Switzerland, Cham (2023). ISBN 978-3-031-44511-8. https://doi.org/10.1007/978-3-031-44511-8_4
7. Dhage, J.S., Gulve, A.K.: Time-distributed layer convolutions with long short-term memory for human activity recognition and result comparison with various machine learning models. Eng. Technol. Appl. Sci. Res. **15**, 23277–23282 (2025). https://doi.org/10.48084/etasr.10499
8. Dosovitskiy, A., et al.: An image is worth 16x16 words. arXiv preprint arXiv:2010.11929 (2020)
9. Falconí Estrada, L.G.: Transfer and ensemble learning models in breast mammogram pathology classification. Master's thesis, Escuela Politécnica Nacional, Quito, Ecuador (2020). https://bibdigital.epn.edu.ec/handle/15000/20853
10. Fan, H., et al.: Multiscale vision transformers. In: Proceedings of the IEEE International Conference on Computer Vision pp. 6804–6815 (2021). https://doi.org/10.1109/ICCV48922.2021.00675, https://arxiv.org/pdf/2104.11227
11. Garia, L.S., Hariharan, M.: Vision transformers for breast cancer classification from thermal images. Lecture Notes in Electrical Engineering 1009 LNEE, pp. 177–185 (2023). https://doi.org/10.1007/978-981-99-0236-1_13, https://link.springer.com/chapter/10.1007/978-981-99-0236-1_13
12. Ji, S., Xu, W., Yang, M., Yu, K.: 3D convolutional neural networks for human action recognition. IEEE Trans. Pattern Anal. Mach. Intell. **35**, 221–231 (2013). https://doi.org/10.1109/TPAMI.2012.59
13. Kennedy, D.A., Lee, T., Seely, D.: A comparative review of thermography as a breast cancer screening technique (2009). https://doi.org/10.1177/1534735408326171

14. Kitchenham, B., Brereton, O.P., Budgen, D., Turner, M., Bailey, J., Linkman, S.: Systematic literature reviews in software engineering-a systematic literature review. Inf. Softw. Technol. **51**(1), 7–15 (2009)
15. Kolesnikov, A., et al.: Big transfer (bit): general visual representation learning. LNCS (including subseries Lecture Notes in Artificial Intelligence and Lecture Notes in Bioinformatics) LNCS, vol. 12350, pp. 491–507 (2019). https://doi.org/10.1007/978-3-030-58558-7_29, https://arxiv.org/pdf/1912.11370
16. Li, Y., Liang, Q., Gan, B., Cui, X.: Action recognition and detection based on deep learning: a comprehensive summary. Comput. Mater. Continua **77**, 1–23 (2023).https://doi.org/10.32604/CMC.2023.042494, https://www.sciencedirect.com/org/science/article/pii/S1546221823001285
17. Litjens, G., et al.: A survey on deep learning in medical image analysis. Med. Image Anal. **42**, 60–88 (2017). https://doi.org/10.1016/J.MEDIA.2017.07.005
18. Liu, D., Wu, B., Li, C., Sun, Z., Zhang, N.: TrenD: a transformer-based encoder-decoder model with adaptive patch embedding for mass segmentation in mammograms. Med. Phys. **50**, 2884–2899 (2023). https://doi.org/10.1002/mp.16216
19. Rakhunde, M.B., Gotarkar, S., Choudhari, S.G.: Thermography as a breast cancer screening technique: a review article. Cureus (2022). https://doi.org/10.7759/cureus.31251
20. Ryan, L., Agaian, S.: Breast cancer detection using infrared thermography: a survey of texture analysis and machine learning approaches. Bioengineering **12**, 639 (2025). https://doi.org/10.3390/BIOENGINEERING12060639, https://pmc.ncbi.nlm.nih.gov/articles/PMC12189745/
21. Selva, J., Johansen, A.S., Escalera, S., Nasrollahi, K., Moeslund, T.B., Clapés, A.: Video transformers: a survey. IEEE Trans. Pattern Anal. Mach. Intell. **45**(11), 12922–12943 (2023). https://doi.org/10.1109/tpami.2023.3243465
22. Silva, L.F., et al.: A new database for breast research with infrared image. J. Med. Imaging Health Inf. **4**, 92–100 (2014). https://doi.org/10.1166/jmihi.2014.1226
23. Su, Y., Liu, Q., Xie, W., Hu, P.: YOLO-LOGO: a transformer-based YOLO segmentation model for breast mass detection and segmentation in digital mammograms. Comput. Methods Programs Biomed. **221** (2022). https://doi.org/10.1016/j.cmpb.2022.106903
24. Tran, D., Wang, H., Torresani, L., Ray, J., Lecun, Y., Paluri, M.: A closer look at spatiotemporal convolutions for action recognition. In: Proceedings of the IEEE Computer Society Conference on Computer Vision and Pattern Recognition, pp. 6450–6459 (11 2017). https://doi.org/10.1109/CVPR.2018.00675, https://arxiv.org/pdf/1711.11248
25. Tsietso, D., Yahya, A., Samikannu, R.: A review on thermal imaging-based breast cancer detection using deep learning. Mob. Inf. Syst. **2022**(1), 8952849 (2022)
26. Vaswani, A., et al.: Attention is all you need. In: Guyon, I., et al. (eds.) Advances in Neural Information Processing Systems, vol. 30. Curran Associates, Inc. (2017). https://proceedings.neurips.cc/paper_files/paper/2017/file/3f5ee243547dee91fbd053c1c4a845aa-Paper.pdf
27. Visual-Lab: Captura de imagens térmicas no huap (rotina diária e protocolos) (2014). https://visual.ic.uff.br/proeng/protocolo.pdf

Multi-view Thermal Breast Imaging for Malignancy Detection: Performance Benchmarking of CNN, Transformer, and Involution Architectures

Mücahit Cihan$^{(\boxtimes)}$ (ID) and Murat Ceylan (ID)

Faculty of Engineering and Natural Sciences, Department of Electrical and Electronics Engineering, Konya Technical University, Konya, Türkiye
{mcihan,mceylan}@ktun.edu.tr

Abstract. Breast cancer screening demands accurate, non-invasive, low-cost tools. Infrared thermography is radiation-free and portable, but its utility hinges on robust computer-aided diagnosis (CAD). We benchmark three deep-learning families for static multi-view breast thermography—CNNs, Transformers, and an involution-based model (HarmonyNet-Lite). Experiments use the Breast Thermography dataset (119 patients; 476 manually segmented ROIs from anterior/oblique views). A compact pipeline performs ROI segmentation, RGB conversion, normalization, resizing, and moderate data augmentation; class imbalance is handled with minority oversampling and class-weighted loss. Evaluation follows patient-stratified five-fold cross-validation. HarmonyNet-Lite yields the best results: accuracy 87.47 ± 2.99%, recall 93.33 ± 2.13%, F1 68.43 ± 8.75%, and precision 54.23 ± 8.94%, indicating high sensitivity with an acceptable trade-off in precision for screening. Among CNNs, ResNet50 is strongest (85.59 ± 3.37%; F1 63.16 ± 3.87%), followed by InceptionV3 (83.38 ± 1.41%; F1 59.99 ± 6.72%), while DenseNet121 lags (79.25 ± 2.98%; F1 52.38 ± 5.62%). Transformer performance is mixed: ViT-Tiny is competitive (84.59 ± 4.23%; F1 59.46 ± 4.68%), whereas Swin-Tiny trails (81.30 ± 2.32%; F1 57.14 ± 4.44%) due to lower precision. Despite using only 0.14 M parameters, HarmonyNet-Lite outperforms heavier CNNs (ResNet50: 23.59 M; InceptionV3: 21.81 M) and lighter Transformers (ViT-Tiny: 2.84 M; Swin-Tiny: 11.78 M), demonstrating that content-adaptive, spatially aware involution operators efficiently capture fine thermal gradients. These findings position HarmonyNet-Lite as a strong, deployable CAD candidate. Future work will pursue multi-center validation, automated segmentation, multi-class labeling, hybrid involution–attention/multimodal models, and controlled GAN-based augmentation to mitigate data scarcity.

Keywords: Breast thermography · Deep learning · HarmonyNet-Lite

1 Introduction

Breast cancer remains one of the most prevalent malignancies and a leading cause of cancer mortality among women worldwide [1]. Early detection is crucial for improving survival rates, yet conventional screening methods present inherent limitations.

© The Author(s), under exclusive license to Springer Nature Switzerland AG 2026
S. T. Kakileti et al. (Eds.): AIIIMA 2025, LNCS 16308, pp. 20–35, 2026.
https://doi.org/10.1007/978-3-032-10990-3_2

Mammography, the current gold standard, may be less effective for women with dense breast tissue and exposes patients to ionizing radiation, causing potential discomfort [2]. These limitations have stimulated increasing interest in infrared thermography as a non-invasive, radiation-free adjunct or alternative for breast cancer screening [3]. Thermal imaging captures infrared heat emissions from the skin; tumors can induce angiogenesis and abnormal blood flow, leading to localized temperature increases detectable on thermograms [3]. This modality is pain-free, cost-effective, portable, and suitable for repeated use—making it particularly attractive for routine monitoring and deployment in resource-constrained settings [1, 3, 4].

In the last decade, the field of medical thermography has experienced a resurgence, driven by advances in artificial intelligence (AI), particularly deep learning, which have significantly improved interpretability and diagnostic accuracy [5, 6]. Recent years have witnessed the emergence of numerous computer-aided diagnosis (CAD) systems leveraging deep neural networks to detect subtle thermal signatures indicative of malignancy. Convolutional Neural Networks (CNNs), known for their ability to learn hierarchical image features, have demonstrated strong performance in breast thermogram classification tasks [7]. For example, Mohamed et al. developed a fully automated pipeline combining U-Net–based breast region segmentation with a custom CNN classifier, achieving over 99% accuracy on the DMR-IR infrared breast dataset [8]. Dharani et al. proposed an Enhanced Deep CNN (EDCNN) with heatmap-based vascularity analysis and fuzzy c-means clustering for Regions of Interest (ROI) extraction, reporting 96.8% sensitivity and 93.7% specificity [9].

While traditional machine learning methods—such as SVM, KNN, decision trees, and LDA—have been applied to thermographic data using handcrafted statistical and texture features, their performance often plateaus due to the complexity of thermal patterns [10]. Deep learning, especially with transfer learning from pretrained models (e.g., ResNet50, DenseNet121, InceptionV3), has proven more effective at extracting discriminative thermal features [6, 11]. Ekici and Jawzal achieved 98.95% accuracy on a breast thermography dataset using a CNN optimized via Bayesian hyperparameter search [12]. Nogales et al. reported that an ensemble of three CNNs focusing on different temperature ranges improved classification accuracy to ~94% [13].

A critical preprocessing step in many thermography-based CAD pipelines is breast region segmentation, aimed at isolating the relevant anatomy and removing background noise [1, 14]. Early works often relied on manual segmentation, especially when dealing with low-contrast thermal boundaries [15]. Manual ROI delineation, while time-consuming, ensures higher precision in focusing the model on diagnostically important regions. More recent studies have explored automatic segmentation using methods like gradient vector flow [16] or deep architectures such as U-Net [8], reducing annotation workload while maintaining ROI accuracy. Nevertheless, manual segmentation remains relevant in research settings, particularly for ensuring data quality and for comparative performance evaluations.

Beyond segmentation, the community has explored data augmentation, oversampling (e.g., MixUp), and multi-view analysis to improve classification robustness [17]. Multi-view thermography—integrating anterior and oblique breast images—has been shown to capture complementary vascular and thermal features, increasing model sensitivity. Additionally, researchers have begun incorporating transformer architectures (e.g., Vision Transformers, Swin Transformer) for their global attention capabilities [18], and experimenting with hybrid CNN-transformer models for medical image classification.

Existing studies emphasize interpretable radiomics, multi-view–to-3D reconstruction, and dynamic IR analysis via 3D CNNs, whereas we evaluate CNN/Transformer/involution models in a static multi-view regime and find involution to be both sensitive and parameter-efficient for local thermal cues [19–21].

An emerging line of research focuses on novel neural operators that can replace or complement convolution. The involution operator, introduced by Li et al. (CVPR 2021) [22], generates spatially adaptive, content-dependent kernels, enabling better spatial specificity with fewer parameters. In the biomedical domain, Cihan et al. proposed HarmonyNet, an involution-based architecture for hyperspectral neonatal health assessment, achieving state-of-the-art performance with high parameter efficiency [23]. This approach offers a compelling direction for thermal imaging, where both fine spatial detail and efficient computation are critical.

In this study, a multi-view thermal breast cancer classification framework is proposed, incorporating CNNs, transformer-based models, and an involution-based HarmonyNet architecture. The dataset includes anterior and oblique thermal views, each manually segmented into left and right ROIs to enhance diagnostic focus. Transfer learning CNNs, transformer models, and HarmonyNet-Lite are systematically compared under consistent evaluation protocols, with performance reported at image-level using fixed and optimal thresholds. The experimental results indicate that HarmonyNet-Lite achieves competitive or superior performance while maintaining computational efficiency, demonstrating its potential for real-world thermal imaging–based screening systems. The overall pipeline of the proposed framework, including data collection, ROI segmentation, preprocessing and augmentation, model training, and evaluation, is illustrated in Fig. 1.

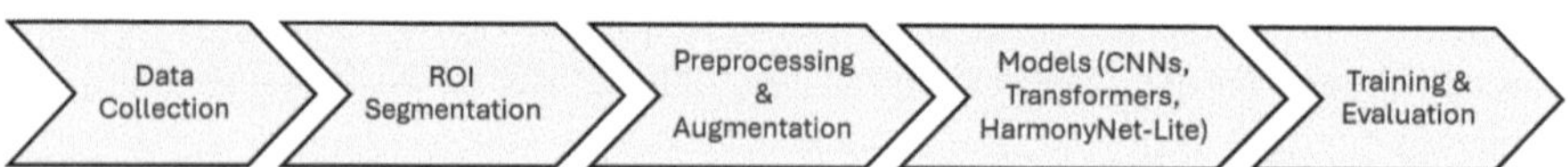

Fig. 1. Workflow pipeline of the proposed study, consisting of data collection, ROI segmentation, preprocessing & augmentation, modeling (CNN, Transformer, HarmonyNet-Lite), and training & evaluation.

2 Materials and Methods

2.1 Dataset and Labeling Procedure

The experimental evaluation was conducted using the publicly available Breast Thermography dataset [24], comprising thermal breast images from 119 individual subjects. For each subject, three distinct anatomical views were provided: Anterior (A), Oblique Left (OL), and Oblique Right (OR). All images were acquired using a FLIR A300 infrared camera under controlled laboratory conditions, maintaining an ambient temperature of 22–24 °C and a relative humidity of approximately 45–50%, in strict accordance with the American Academy of Thermology (AAT) imaging protocol [24].

Each thermal image underwent manual segmentation to isolate the breast region from the surrounding background with high precision. In the anterior view, although only a single image per subject was available, it was manually divided into two distinct ROIs: Anterior Left (AL), corresponding to the subject's left breast, and Anterior Right (AR), corresponding to the right breast. This ensured that each breast was analyzed independently, preserved spatial symmetry, and enabled side-specific diagnostic classification.

For oblique views, a single ROI was extracted from each image, corresponding to the visible breast in that orientation (OL for the left breast, OR for the right breast). Segmentation involved detailed visual inspection to delineate anatomical breast boundaries, followed by cropping or masking to remove non-relevant anatomical structures and thermal noise (e.g., shoulders, neck, or clothing edges). ROI dimensions and spatial alignment were standardized across subjects to maintain geometric consistency during training and evaluation.

Manual segmentation was deemed essential due to the subtle temperature gradients characteristic of infrared thermography, where even small deviations in ROI boundaries can significantly alter thermal distribution patterns and adversely impact classification accuracy. By ensuring consistent and anatomically precise ROI extraction, segmentation reduced inter-sample variability and improved the reliability of subsequent diagnostic modeling (Fig. 2). Manual ROIs were applied to ensure consistent lesion coverage in lieu of widely adopted thermography segmentation tools, and the attendant limitations in scalability are acknowledged as a practical constraint.

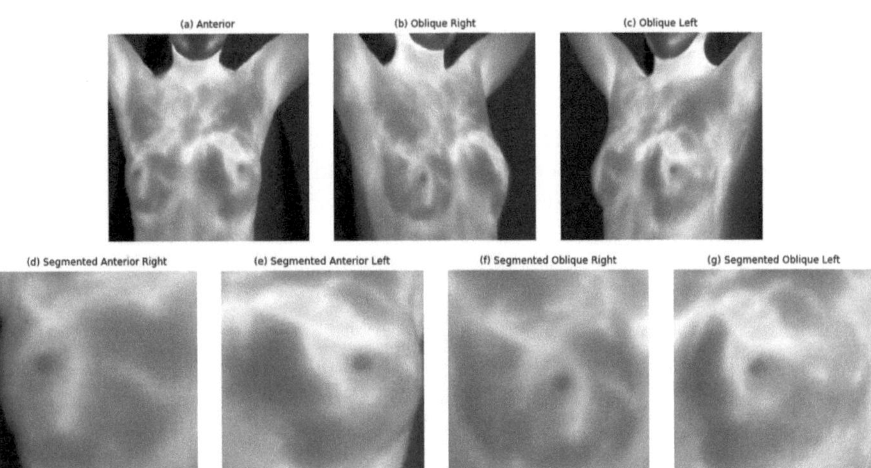

Fig. 2. Examples of original and manually segmented thermal breast images: (a) Anterior, (b) Oblique Right, (c) Oblique Left, (d) Segmented Anterior Right, (e) Segmented Anterior Left, (f) Segmented Oblique Right, (g) Segmented Oblique Left.

Diagnostic labeling was performed independently for each breast. In anterior views, the AL and AR ROIs inherited the diagnostic label assigned to their respective breast by the medical ground truth. In oblique views, labels were assigned directly based on the breast depicted (OL for left, OR for right).

Three diagnostic categories were defined: Normal (N), Probably Benign (PB), and Probably Malignant (PM). For binary classification tasks, N and PB were merged into the benign class (label = 0), while PM constituted the malignant class (label = 1).

In total, the dataset comprised 476 thermal ROIs (AL, AR, OL, OR) derived from 119 subjects. A comprehensive summary of dataset statistics, including the total number of subjects, ROI counts, image modalities, class distribution, and problem definition, is provided in Table 1. All subject-related information—including file paths for anterior and oblique images, diagnostic labels for each breast, and optional metadata (age, weight, height, and body temperature)—was compiled into a structured CSV file. This file served as a standardized reference for all data loading, preprocessing, and training operations.

Table 1. Dataset Overview.

Feature	Value
Total Patients	119
Total ROIs	476
Image Types	Anterior (A), Oblique Left (OL), Oblique Right (OR)
ROI Categories	AL (Anterior-Left), AR (Anterior-Right), OL (Oblique-Left), OR (Oblique-Right)
Class Distribution	Normal (N): 234, Probable Benign (PB): 172, Probable Malignant (PM): 70
Problem Definition	Binary classification (Benign vs. Malignant)

To prevent data leakage and ensure robust evaluation, we adopted patient-level, stratified five-fold cross-validation ($K = 5$). In each fold, all images from a given patient were assigned exclusively to either training or validation, and the malignant–benign ratio was preserved via stratification. The inherent class imbalance, typical in medical imaging datasets, was addressed by combining minority-class oversampling with MixUp augmentation ($\alpha = 0.4$), which improved generalization and mitigated overfitting.

This structured segmentation and labeling pipeline enabled consistent image-level evaluation, ensuring that all predictions were derived from well-defined ROIs and thereby enhancing the reliability and reproducibility of the experimental results.

2.2 Preprocessing and Data Augmentation

To ensure reliable analysis, all thermal images were manually segmented into four clinically relevant ROIs, corresponding to the AL, AR, OL, and OR views. This segmentation step removed background artifacts such as the neck, shoulders, and clothing, thereby restricting the learning process to diagnostically meaningful breast tissue patterns.

Each ROI was converted from grayscale to RGB to ensure compatibility with pretrained architectures, normalized to the [0,1] range for stable optimization, and resized to a fixed resolution of 224 × 224 pixels using bilinear interpolation, which preserved edge sharpness and minimized distortions in thermal intensity distributions.

To improve generalization, online data augmentation was applied during training using TensorFlow's *tf.image* and Keras layers. The strategies included geometric transformations (rotations ± 20°, translations ± 20%, zoom − 20% to + 25%, horizontal flips at 50% probability), resizing to 256 × 256 followed by center cropping to 224 × 224, and cutout masking (35% probability, −25% area) to promote reliance on distributed cues. Additionally, Gaussian noise ($\mu = 0$, $\sigma = 0.01$) was injected to simulate sensor variability, and MixUp augmentation (λ from Beta distribution, $\alpha = 0.4$) was used to increase sample diversity and smooth decision boundaries. Figure 3 illustrates these transformations, showing rotation, translation, zoom, flip, cropping, cutout, Gaussian noise, and MixUp.

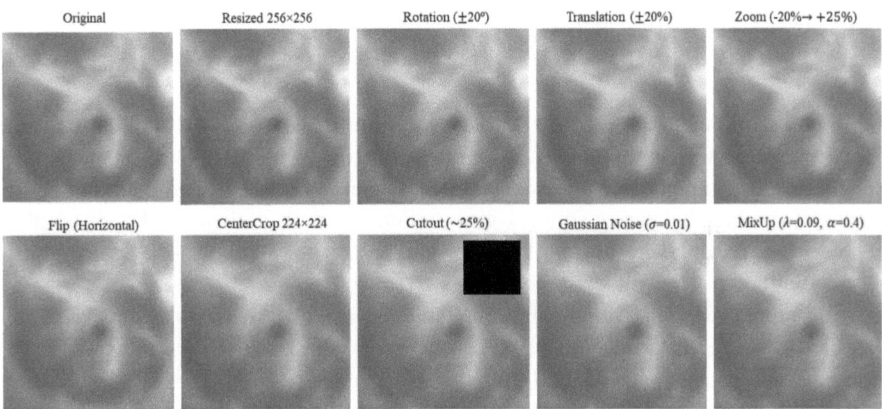

Fig. 3. Illustration of augmentation diversity. The same thermal image was transformed using different augmentation strategies: resizing to 256 × 256, rotation (±20°), translation (±20%), zooming (−20% to + 25%), horizontal flip (p = 0.5), center cropping (224 × 224), cutout (−25%), gaussian noise (σ=0.01) and MixUp (λ=0.09, α=0.4).

The complete preprocessing and augmentation configuration, including normalization, geometric transformations, cutout masking, noise injection, and MixUp strategies, is summarized in Table 2. Augmentations were chosen to reflect thermal acquisition variability (e.g., conservative intensity scaling and noise), and color-based transforms were avoided because thermograms are inherently single-channel.

Table 2. Preprocessing and Augmentation Parameters.

Operation	Parameters
Normalization	Rescaling to [0,1]
Resized	256×256
Rotation	$\pm 20°$
Zoom	$-20\%\ /\ +25\%$
Translation	Horizontal/Vertical $\pm 20\%$
Horizontal Flip	Probability $= 0.5$
Center Crop	224×224
Cutout Masking	Probability $= 0.35$, area $\approx 25\%$
Gaussian Noise	$\mu = 0$, $\sigma = 0.01$
MixUp	$\lambda = 0.09$, $\alpha = 0.4$ (applied to minority class only)

For validation and testing, only resizing and normalization were applied to ensure consistent evaluation without stochastic transformations.

Given the inherent class imbalance between benign (N + PB) and malignant (PM) cases, a two-stage balancing strategy was implemented: (i) patient-level oversampling of malignant cases to avoid data leakage, and (ii) class-weighted loss adjustment to penalize minority-class misclassification more strongly.

All preprocessing and augmentation steps were embedded into an optimized tf.data pipeline featuring shuffling, mini-batch generation (batch size $= 16$), parallel CPU-based mapping, and prefetching (*AUTOTUNE*) to overlap data loading with GPU execution. Random seeds across Python, NumPy, and TensorFlow were fixed (*SEED = 42*) to ensure reproducibility. For evaluation, we used patient-wise, stratified five-fold cross-validation to maintain class balance and prevent subject overlap across folds; training and validation splits were regenerated per fold under the same pipeline.

2.3 Employed Deep Learning Architectures

To comprehensively assess the efficacy of different deep learning paradigms in multi-view breast thermography, three families of models were implemented: Convolutional Neural Networks, Transformer-based architectures, and an involution-based HarmonyNet-Lite framework. All models were trained and evaluated under identical preprocessing, augmentation, and stratified patient-wise splitting protocols to ensure a fair comparison.

Convolutional Neural Networks (CNNs). Three widely adopted CNN backbones, pretrained on ImageNet and fine-tuned for the thermal ROI dataset, were investigated:

- DenseNet121: Exploits dense connectivity, where each layer receives feature maps from all preceding layers. This improves gradient flow, reduces parameter redundancy, and encourages feature reuse.

- InceptionV3: Leverages inception modules with multi-scale convolutional kernels (1×1, 3×3, 5×5), enabling efficient multi-resolution feature extraction while maintaining computational efficiency through factorized convolutions.
- ResNet50: Introduces residual skip connections to alleviate vanishing gradients, allowing stable training of deep networks with balanced accuracy and efficiency.

In all CNNs, the classification head was redesigned with global average pooling and a fully connected dense layer with sigmoid activation, tailored for binary classification tasks.

Transformer-Based Architectures. Two transformer-based vision models were adapted to process the 224×224 thermal ROIs:

- Swin-Tiny Transformer: Employs hierarchical shifted-window self-attention, enabling local-to-global representation learning while avoiding the quadratic cost of global attention.
- Vision Transformer (ViT-Tiny): Operates on non-overlapping image patches, modeling long-range dependencies with multi-head global attention while maintaining positional embeddings to preserve spatial context.

These models are particularly suited for capturing global thermal distribution patterns, complementing the local feature focus of CNNs.

HarmonyNet-Lite (Involution-Based Architecture). HarmonyNet-Lite represents a lightweight adaptation of the original HarmonyNet framework, specifically tailored for thermal ROI classification (AL/AR, OL/OR). Its design emphasizes computational efficiency and reduced parameterization, making it suitable for limited hardware environments and real-time applications.

Unlike conventional convolution, involution layers generate spatially adaptive kernels conditioned on local feature content, allowing fine-grained modeling of subtle thermal variations [25]. The architecture follows a compact hierarchical structure:

1. Stem Layer: Low-width convolutional block for initial feature extraction.
2. Involution Residual Blocks: Sequential stages with progressive channel expansion and spatial downsampling. Smaller receptive fields (3×3) capture local texture details, while deeper blocks with larger kernels (5×5, 7×7) extend contextual awareness.
3. Classification Head: Global average pooling, dropout, and a dense layer with sigmoid activation for binary decision-making.

To enhance generalization, stochastic depth (DropPath), dropout regularization, and batch normalization were integrated across stages. Training was performed with the Adam optimizer and binary cross-entropy loss, with performance tracked using Accuracy, Precision, and Recall. Learning rate scheduling and early stopping based on validation AUC ensured stable convergence.

By leveraging spatially adaptive filtering, HarmonyNet-Lite was designed to suppress irrelevant heat distributions (e.g., from shoulders or neck) and focus on diagnostically relevant temperature gradients. Within the experimental protocol, HarmonyNet-Lite was evaluated alongside CNN-based (DenseNet, InceptionV3, ResNet50) and Transformer-based (ViT-Tiny, Swin-Tiny) models under identical training and validation conditions. These baselines represent canonical CNN and Transformer families; HarmonyNet-Lite is compared not only in accuracy but also by parameter count and compute cost, while a broader exploration of lightweight variants is deferred to future work.

2.4 Training and Evaluation Metrics

All deep learning models, including classical CNN-based architectures (DenseNet121, InceptionV3, ResNet50), Transformer-based architectures (ViT-Tiny, Swin-Tiny) and the proposed HarmonyNet-Lite, were trained under a consistent experimental protocol to ensure fair comparison.

Training was conducted under patient-level, stratified five-fold cross-validation to avoid data leakage, with each fold evaluating on patients unseen during training. Stratification was performed according to the presence of malignant cases to preserve the malignant-to-benign ratio within each fold. This protocol ensures that reported metrics reflect mean $\pm$ SD across folds, rather than a single fixed 80/20 hold-out split.

Given the inherent class imbalance in the dataset, two complementary strategies were applied:

- Random oversampling of the minority class at the training set level to equalize class distributions.
- MixUp augmentation ($\alpha = 0.4$) applied to minority class samples during training to promote smoother decision boundaries and reduce overfitting.

For CNN-based models, data augmentation included random rotation ($\pm 20°$), translation (up to 20%), zooming (-20% to $+25\%$), and horizontal flipping. Random cutout masking and Gaussian noise addition were also applied. All augmentations were performed online during training to increase data variability. RGB-based pretrained CNNs received equivalent geometric augmentations adapted to their input size requirements (224×224 for ResNet/DenseNet/ViT-Tiny, and 299×299 for InceptionV3).

Optimization was performed using the Adam optimizer with binary cross-entropy loss. An early stopping mechanism monitored the validation loss with a patience of six epochs, restoring the best model weights upon termination. A *ReduceLROnPlateau* scheduler was employed to adaptively decrease the learning rate when validation performance plateaued. HarmonyNet-Lite underwent an additional fine-tuning phase with a reduced learning rate ($1e-4$).

Model performance was evaluated at the image level, with predictions generated independently for each ROI. We report Accuracy, Precision, Recall, F1-score (F1), and the number of trainable parameters, using a fixed decision threshold of 0.50. To mitigate overfitting on the 119-patient cohort, we employed patient-level, stratified five-fold (K = 5) cross-validation. Performance is presented as fold-wise mean $\pm$ SD across the five folds.

3 Experimental Setup

3.1 Hardware and Software Environment

All experiments were performed on a dedicated workstation configured with an NVIDIA GeForce GTX 1080 Ti GPU with 11 GB GDDR5X memory, an Intel® Xeon® E5–2680 processor running at 2.70 GHz with 32 cores, and 64 GB DDR4 RAM. The system was equipped with a 2 TB NVMe SSD for high-speed storage and operated under Microsoft Windows 10 Pro (64-bit).

Model development, training, and evaluation were implemented in Python 3.9 using the TensorFlow 2.10.1 deep learning framework with Keras 2.10.0 as the high-level API. Image preprocessing and augmentation were performed using TensorFlow's tf.image module and Keras preprocessing layers, while NumPy (1.23.5) supported array manipulation and Pandas (1.4.4) facilitated dataset management through CSV-based operations.

For visualization and statistical analysis, Matplotlib (3.6.3) was used to plot training curves and metrics; Scikit-learn (1.1.3) computed the evaluation metrics. To ensure full reproducibility, all experiments were executed in a controlled environment with fixed random seeds across Python, NumPy, and TensorFlow.

3.2 Training Parameters

All deep learning models, including CNN-based architectures (DenseNet121, InceptionV3, ResNet50), Transformer-based architectures (ViT-Tiny, Swin-Tiny), and the involution-based HarmonyNet-Lite, were trained under a unified optimization and hyper-parameter protocol. The training employed the Adam optimizer ($\beta_1 = 0.9$, $\beta_2 = 0.999$, $\varepsilon = 1 \times 10^{-7}$) with binary cross-entropy loss, where class weighting was optionally introduced to address imbalance. The initial learning rate was set to 1×10^{-3} during transfer learning and reduced to 1×10^{-4} in the fine-tuning stage, while learning rate scheduling was managed by a ReduceLROnPlateau strategy (factor 0.2, patience 3, minimum LR 1×10^{-6}). Early stopping was applied by monitoring the validation ROC-AUC with a patience of six epochs, ensuring that the best-performing weights were restored. Training was carried out with a batch size of 16 for up to 100 epochs, subject to early termination when convergence criteria were met. Pretrained ImageNet weights were used for CNN backbones, whereas Transformer models and HarmonyNet-Lite were initialized randomly.

For transfer learning models, a two-phase strategy was adopted. In the first phase, only the classification head was trained while keeping the backbone frozen. In the second phase, the top layers of the backbone were unfrozen and fine-tuned with a reduced learning rate. HarmonyNet-Lite underwent an additional fine-tuning stage with a learning rate of 1×10^{-4} after initial convergence, further refining feature representations and enhancing model stability. To promote generalization, online data augmentation (as described in Sect. 2.2) was consistently applied, including random rotations, translations, zooming, horizontal flips, cutout masking, and Gaussian noise injection for grayscale inputs.

4 Results and Discussion

All models were evaluated on ROI-level predictions (AL, AR, OL, OR). Table 3 reports five-fold, patient-stratified cross-validation results as mean $\pm$ SD, and Fig. 4 visualizes these metrics with error bars denoting fold-to-fold variability. Across baselines, the proposed HarmonyNet-Lite achieved the highest accuracy (87.47 $\pm$ 2.99%), recall (93.33 $\pm$ 2.13%), and F1-score (68.43 $\pm$ 8.75%), while maintaining a balanced precision (54.23 $\pm$ 8.94%). This consistently high recall is particularly important for screening, reflecting a lower propensity to miss malignant cases; the moderate precision indicates the usual recall-oriented trade-off toward a small number of additional false positives—an acceptable compromise in early-detection workflows.

Among CNN backbones, DenseNet121 was the weakest performer (79.25 $\pm$ 2.98% accuracy; 52.38 $\pm$ 5.62% F1), limited by a low precision profile (39.33 $\pm$ 5.72%) despite a mid-range recall (79.39 $\pm$ 2.59%). InceptionV3 offered a more balanced outcome (83.38 $\pm$ 1.41% accuracy; 59.99 $\pm$ 6.72% F1) with 46.00 $\pm$ 6.82% precision and 85.52 $\pm$ 3.12% recall. The strongest CNN baseline, ResNet50, reached 85.59 $\pm$ 3.37% accuracy and 63.16 $\pm$ 3.87% F1 with 50.13 $\pm$ 3.25% precision and 85.82 $\pm$ 2.05% recall—approaching HarmonyNet-Lite but still trailing it on all headline metrics.

Table 3. Five-fold cross-validation performance comparison of CNN-, Transformer-, and Involution-based models for breast thermography classification (mean $\pm$ SD).

Model	Accuracy (%)	Precision (%)	Recall (%)	F1-score (%)	Params (M)
HarmonyNet-Lite	**87.47 $\pm$ 2.99**	**54.23 $\pm$ 8.94**	**93.33 $\pm$ 2.13**	**68.43 $\pm$ 8.75**	**0.14**
ResNet50	85.59 $\pm$ 3.37	50.13 $\pm$ 3.25	85.82 $\pm$ 2.05	63.16 $\pm$ 3.87	23.59
ViT-Tiny	84.59 $\pm$ 4.23	48.00 $\pm$ 4.47	79.37 $\pm$ 3.14	59.46 $\pm$ 4.68	2.84
InceptionV3	83.38 $\pm$ 1.41	46.00 $\pm$ 6.82	85.52 $\pm$ 3.12	59.99 $\pm$ 6.72	21.81
Swin-Tiny	81.30 $\pm$ 2.32	42.57 $\pm$ 4.33	85.21 $\pm$ 3.21	57.14 $\pm$ 4.44	11.78
DenseNet121	79.25 $\pm$ 2.98	39.33 $\pm$ 5.72	79.39 $\pm$ 2.59	52.38 $\pm$ 5.62	7.04

Transformer results were mixed. ViT-Tiny was competitive (84.59 $\pm$ 4.23% accuracy; 59.46 $\pm$ 4.68% F1) with 48.00 $\pm$ 4.47% precision and 79.37 $\pm$ 3.14% recall, whereas Swin-Tiny lagged (81.30 $\pm$ 2.32% accuracy; 57.14 $\pm$ 4.44% F1), driven by reduced precision (42.57 $\pm$ 4.33%) despite relatively high recall (85.21 $\pm$ 3.21%). On this modest-sized cohort, Transformers appear more sensitive to optimization/regularization under low-contrast thermal cues; local operators (CNNs/involution) seem to offer a more stable inductive bias.

Parameter efficiency is a key differentiator. HarmonyNet-Lite uses only 0.14 M parameters yet surpasses heavier CNNs (ResNet50 23.59 M, InceptionV3 21.81 M) and even lighter Transformers (ViT-Tiny 2.84 M, Swin-Tiny 11.78 M). DenseNet121 (7.04 M) presents moderate complexity but weaker accuracy/F1. This efficiency-to-accuracy ratio positions HarmonyNet-Lite well for portable CAD scenarios and resource-constrained deployments.

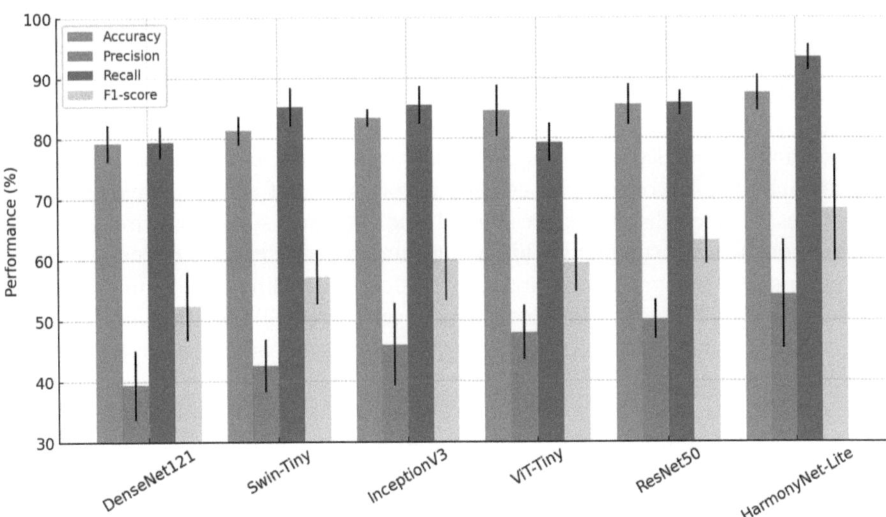

Fig. 4. Mean ± SD performance of CNN-, Transformer-, and HarmonyNet-Lite–based models on breast thermography classification. Error bars denote variability across five folds.

The fold-level results in Table 4 corroborate the aggregate findings. HarmonyNet-Lite is top in all five folds (range 84.48–90.46%, SD 2.99%). ResNet50 is consistently second (82.22–88.96%, SD 3.37%). ViT-Tiny is competitive but shows the largest variability (80.36–88.82%, SD 4.23%). InceptionV3 is the most stable overall (range 81.97–84.79%, SD 1.41%) yet trails in absolute accuracy. Swin-Tiny (78.98–83.62%, SD 2.32%) and DenseNet121 (76.27–82.23%, SD 2.98%) remain lower across folds. These per-fold patterns align with the precision/recall trade-offs noted above (e.g., HarmonyNet-Lite's fewer false negatives vs. Swin-Tiny/DenseNet121's higher false positives).

Table 4. Five-fold patient-stratified cross-validation accuracy per fold.

Model	Fold1	Fold2	Fold3	Fold4	Fold5	Mean ± SD
HarmonyNet-Lite	84.48	86.12	87.93	90.46	88.36	87.47 ± 2.99
ResNet50	82.22	84.12	86.11	88.96	86.54	85.59 ± 3.37
ViT-Tiny	80.36	83.73	85.02	88.82	85.02	84.59 ± 4.23
InceptionV3	81.97	83.17	83.53	84.79	83.44	83.38 ± 1.41
Swin-Tiny	78.98	80.41	81.31	83.62	82.18	81.30 ± 2.32
DenseNet121	76.27	78.29	79.72	82.23	79.74	79.25 ± 2.98

From a literature standpoint, prior studies in breast thermography have largely emphasized interpretable radiomics or modeled 3D geometry/dynamic IR streams. In contrast, our static multi-view benchmark shows that involution-based, content-adaptive spatial filtering captures fine local thermal gradients with substantially fewer parameters, yielding strong recall and competitive overall accuracy—properties that are valuable in screening pipelines and edge settings [4, 19, 20].

Spatially adaptive kernels in involution are hypothesized to preserve subtle, local thermal gradients more effectively than position-shared convolutions or globally focused self-attention—an inductive bias aligned with the small, low-contrast cues prevalent in thermography. Unlike conventional convolution or global self-attention, this mechanism emphasizes localized, discriminative thermal gradients while suppressing irrelevant heat patterns from non-breast regions. This architectural property enables HarmonyNet-Lite to capture subtle malignant variations more effectively, while maintaining computational efficiency with fewer parameters than larger CNN or Transformer models. Such characteristics make it suitable for real-world deployment, particularly in portable and resource-limited settings.

These findings should be interpreted with caveats: a single-center, modest cohort (119 patients; 476 ROIs), reliance on manual ROIs that aid data quality yet limit scalability, and a binary label scheme reducing clinical granularity. Within this scope, HarmonyNet-Lite's high recall and 0.14 M-parameter footprint indicate a favorable sensitivity–efficiency trade-off for screening-oriented CAD.

5 Conclusion and Future Work

This study presented HarmonyNet-Lite, an involution-based deep learning architecture tailored for breast thermography classification. Using patient-stratified five-fold cross-validation, HarmonyNet-Lite achieved the highest accuracy ($87.47 \pm 2.99\%$), recall ($93.33 \pm 2.13\%$), and F1-score ($68.43 \pm 8.75\%$), while maintaining balanced precision ($54.23 \pm 8.94\%$). These results indicate a strong ability to detect malignant cases (low false-negative tendency) with an acceptable trade-off of additional false positives for screening workflows. In terms of compute, HarmonyNet-Lite is highly parameter-efficient (0.14 M parameters), yet it outperforms much heavier CNNs (ResNet50: 23.59 M; InceptionV3: 21.81 M) and even lighter Transformers (ViT-Tiny: 2.84 M; Swin-Tiny: 11.78 M). Taken together, the findings support the value of content-adaptive, spatially aware (involution) filtering for capturing subtle, local thermal gradients that conventional convolution or global self-attention may overlook, delivering superior sensitivity at a fraction of the model size.

These results should be interpreted considering several limitations: the single-center, modest cohort (119 patients; 476 ROIs), reliance on manual ROI segmentation that improves data quality yet limits scalability, and the use of a binary label schema (benign vs. malignant) that reduces clinical granularity. In addition, precision remains moderate across models (including HarmonyNet-Lite), implying a non-trivial rate of false-positive callbacks in a screening setting. These factors define the scope of the reported results.

Building on these results, we will (i) pursue multi-center external validation; (ii) replace manual ROIs with automated segmentation (e.g., U-Net) for deployable workflows; (iii) extend to multi-class labels (normal/benign/malignant) to increase clinical utility; (iv) explore hybrid involution–attention architectures and multimodal fusion (thermography + clinical metadata) to boost robustness and interpretability; and (v) mitigate data scarcity via controlled GAN-based augmentation alongside MixUp/oversampling, informed by recent generative studies [26, 27].

In conclusion, HarmonyNet-Lite shows that lightweight, spatially adaptive models can deliver reliable, high-recall performance in breast thermography with excellent computational efficiency. With continued progress on dataset scale, automated segmentation, and hybrid model design, the framework is well-positioned for integration into CAD systems to support early detection and screening in both clinical and resource-constrained environments.

Disclosure of Interests. The authors have no competing interests to declare that are relevant to the content of this article.

References

1. American Cancer Society: Breast Cancer Facts & Figures 2024–2025. American Cancer Society, Atlanta (2024)
2. Hendrick, R.E.: Radiation doses and risks in breast screening. J. Breast Imaging **2**(3), 188–200 (2020). https://doi.org/10.1093/jbi/wbaa016
3. Lahiri, B.B., Bagavathiappan, S., Jayakumar, T., Philip, J.: Medical applications of infrared thermography: a review. Infrared Phys. Technol. **55**(4), 221–235 (2012). https://doi.org/10.1016/j.infrared.2012.03.007
4. Brioschi, G.C., Brioschi, M.L., Dalmaso Neto, C., O'Young, B.: The socioeconomic impact of artificial intelligence applications in diagnostic medical thermography: a comparative analysis with mammography in breast cancer detection and other diseases early detection. In: MICCAI Workshop on Artificial Intelligence over Infrared Images for Medical Applications (AIIIMA), pp. 1–31. Springer, Cham (2023)
5. Nowakowski, A.Z., Kaczmarek, M.: Artificial intelligence in IR thermal imaging and sensing for medical applications. Sensors **25**(3), 891 (2025). https://doi.org/10.3390/s25030891
6. Chantasartrassamee, P., Ongphiphadhanakul, B., Suvikapakornkul, R., Binsirawanich, P., Sriphrapradang, C.: Artificial intelligence-enhanced infrared thermography as a diagnostic tool for thyroid malignancy detection. Ann. Med. **56**(1), 2425826 (2024). https://doi.org/10.1080/07853890.2024.2425826
7. Al Husaini, M.A.S., Habaebi, M.H., Hameed, S.A., Islam, M.R., Gunawan, T.S.: A systematic review of breast cancer detection using thermography and neural networks. IEEE Access **8**, 208922–208937 (2020). https://doi.org/10.1109/ACCESS.2020.3038817
8. Mohamed, E.A., Rashed, E.A., Gaber, T., Karam, O.: Deep learning model for fully automated breast cancer detection system from thermograms. PLoS ONE **17**(1), e0262349 (2022). https://doi.org/10.1371/journal.pone.0262349
9. Dharani, N.P., Immadi, I.G., Narayana, M.V.: Enhanced deep learning model for diagnosing breast cancer using thermal images. Soft. Comput. **28**(13), 8423–8434 (2024). https://doi.org/10.1007/s00500-024-09742-8
10. Khodadadi, H., Nazem, S.: Improving cancer detection through computer-aided diagnosis: a comprehensive analysis of nonlinear and texture features in breast thermograms. PLoS ONE **20**(5), e0322934 (2025). https://doi.org/10.1371/journal.pone.0322934
11. Al-Nasr, A.S., Aref, M., Sabry, Y.M.: A comparative study of AI-driven approaches for breast malignance detection exploiting infrared thermography. In: 2025 15th International Conference Electrical Engineering (ICEENG), pp. 1–5. IEEE, New York (2025). https://doi.org/10.1109/ICEENG64546.2025.11031310
12. Ekici, S., Jawzal, H.: Breast cancer diagnosis using thermography and convolutional neural networks. Med. Hypotheses **137**, 109542 (2020). https://doi.org/10.1016/j.mehy.2019.109542
13. Nogales, A., Perez-Lara, F., García-Tejedor, Á.J.: Enhancing breast cancer diagnosis with deep learning and evolutionary algorithms: a comparison of approaches using different thermographic imaging treatments. Multimed. Tools Appl. **83**(14), 42955–42971 (2024). https://doi.org/10.1007/s11042-023-17281-x

14. Singh, J., Arora, A.S.: Automated approaches for ROIs extraction in medical thermography: a review and future directions. Multimed. Tools Appl. **79**(21), 15273–15296 (2020). https://doi.org/10.1007/s11042-018-7113-z

15. Urrea, C., Vélez, M.: Advances in deep learning for semantic segmentation of low-contrast images: a systematic review of methods, challenges, and future directions. Sensors **25**(7), 2043 (2025). https://doi.org/10.3390/s25072043

16. Tello-Mijares, S., Woo, F., Flores, F.: Breast cancer identification via thermography image segmentation with a gradient vector flow and a convolutional neural network. J. Healthc. Eng. **2019**(1), 9807619 (2019). https://doi.org/10.1155/2019/9807619

17. Jin, X., et al.: A survey on mixup augmentations and beyond. *arXiv preprint* arXiv:2409.05202 (2024). https://doi.org/10.48550/arXiv.2409.05202

18. Vaswani, A., et al.: Attention is all you need. In: Advances in Neural Information Processing Systems (NeurIPS), vol. 30. Curran Associates Inc., Red Hook (2017)

19. Baffa, M.D.F.O., Conci, A.: Radiomics for breast IR-Imaging classification. In: Kakileti, S.T., et al. Artificial Intelligence over Infrared Images for Medical Applications and Medical Image Assisted Biomarker Discovery. MIABID AIIIMA 2022 2022. Lecture Notes in Computer Science, vol. 13602, pp. 10–19. Springer, Cham (2022). https://doi.org/10.1007/978-3-031-19660-7_2

20. Saha, A.P., Kakileti, S.T., Dedhiya, R., Manjunath, G.: 3D-BreastNet: a self-supervised deep learning network for reconstruction of 3D breast surface from 2D thermal images. In: Kakileti, S.T., Manjunath, G., Schwartz, R.G., Frangi, A.F. (eds.) Artificial Intelligence over Infrared Images for Medical Applications. AIIIMA 2023. Lecture Notes in Computer Science, vol. 14298, pp. 32–44. Springer, Cham (2023). https://doi.org/10.1007/978-3-031-44511-8_2

21. de Freitas Oliveira Baffa, M., Neves, T.G.Z., Tulha, C.N., Conci, A.: 3D-CNN for breast cancer detection on angular IR images. In: Kakileti, S.T., Manjunath, G., Schwartz, R.G., Ng, E.Y.K. (eds) Artificial Intelligence over Infrared Images for Medical Applications. AIIIMA 2024. Lecture Notes in Computer Science, vol. 15279, pp. 57–68. Springer, Cham (2025). https://doi.org/10.1007/978-3-031-76584-1_6

22. Li, D., et al.: Involution: Inverting the inherence of convolution for visual recognition. In: Proc. IEEE/CVF Conf. Computer Vision and Pattern Recognition (CVPR), pp. 12321–12330. IEEE, New York (2021)

23. Cihan, M., Ceylan, M., Konak, M., Soylu, H.: Involution-based HarmonyNet: an efficient hyperspectral imaging model for automatic detection of neonatal health status. Biomed. Signal Process. Control **100**, 106982 (2025). https://doi.org/10.1016/j.bspc.2024.106982

24. Rodriguez-Guerrero, S., et al.: Breast thermography (Version 3). Mendeley Data (2024). https://doi.org/10.17632/mhrt4svjxc.3

25. Cihan, M., Ceylan, M.: HybridCISN: Integrating 2D/3D convolutions and involutions with hyperspectral imaging and blood biomarkers for neonatal disease detection. Comput. Electr. Eng. **123**, 110193 (2025). https://doi.org/10.1016/j.compeleceng.2025.110193

26. Govindaraju, B., Kakileti, S.T.: Generative artificial intelligence approaches for synthesizing high-fidelity breast thermal images. In: Kakileti, S.T., Manjunath, G., Schwartz, R.G., Ng, E.Y.K. (eds.) Artificial Intelligence over Infrared Images for Medical Applications. AIIIMA 2024. Lecture Notes in Computer Science, vol. 15279, pp. 33–43. Springer, Cham (2024). https://doi.org/10.1007/978-3-031-76584-1_4

27. Silva, C.E.C., Conci, A.: About the validity of using DCGANs for data augmentation in breast thermography segmentation. In: Kakileti, S.T., Manjunath, G., Schwartz, R.G., Ng, E.Y.K. (eds.) Artificial Intelligence over Infrared Images for Medical Applications. AIIIMA 2024. Lecture Notes in Computer Science, vol. 15279, pp. 44–56. Springer, Cham (2024). https://doi.org/10.1007/978-3-031-76584-1_5

A Shapley Value-Based Gated Feature Fusion of Multi-branch Deep Learning Framework for Breast Cancer Screening of Thermal Images

Jotiraditya Banerjee[1(✉)], Rajdeep Pal[2], and Ram Sarkar[3]

[1] Department of Power Engineering, Jadavpur University, Kolkata, India
`joti.ban.2710@gmail.com`
[2] Department of Artificial Intelligence and Machine Learning, St. Thomas College of Engineering and Technology, Kolkata, India
[3] Department of Computer Science and Engineering, Jadavpur University, Kolkata, India

Abstract. Breast cancer detection using thermal imaging presents a non-invasive, cost-effective screening alternative, especially suited for resource-limited environments. This paper proposes a multi-branch hybrid deep learning framework that simultaneously processes spatial thermograms and their frequency-domain Power Spectral Density (PSD) representations. A novel Shapley-value based Gated Fusion mechanism adaptively integrates complementary features extracted via a lightweight MobileNetV3Small backbone, enhancing the discriminative capability of the overall model, while maintaining computational efficiency. The model is trained and evaluated on the publicly available DMR-IR dataset of thermal breast images. Experimental results demonstrate superior diagnostic accuracy compared to prior state-of-the-art approaches, highlighting the efficacy of joint spatial-frequency feature learning for breast cancer classification. The codes for this work are available at : https://github.com/Kylian07/Shaply-based-Gated-Fusion.

Keywords: Breast Cancer · Thermal Image · Shapley Value · Feature Fusion · Power Spectral Density · Frequency Domain · DMR-IR Dataset

1 Introduction

The most common disease among women worldwide is breast cancer, which continues to be a serious public health concern. The World Health Organization (WHO) reports that in 2020 alone, there were around 685,000 fatalities, and over 2.3 million new cases of breast cancer [1]. Since survival rates can increase to 96% if treatment is started prior to metastasis, early identification is essential [2]. Despite their widespread use for screening, conventional imaging techniques

S. T. Kakileti et al. (Eds.): AIIIMA 2025, LNCS 16308, pp. 36–49, 2026.
https://doi.org/10.1007/978-3-032-10990-3_3

such as mammography, ultrasound, and magnetic resonance imaging (MRI) have drawbacks such as low diagnostic sensitivity in dense breast tissue, high cost, ionizing radiation exposure, and restricted accessibility in under-resourced areas [3,4]. Despite being the gold-standard confirming test, biopsy is not suitable for routine large-scale screening due to its invasiveness and procedural complexity [5].

The non-invasive, radiation-free, portable, and economical benefits of infrared thermography have made it a Promising Supplementary modality. It functions by recording heat patterns from skin's surface, which can reveal physiological alterations caused by tumors, including aberrant vascularization and angiogenesis [6,7]. Because of these characteristics, thermography is particularly appealing for application in low-resource environments and in repetitive screenings. Its efficacy, however, is largely dependent on reliable computer models that are able to distinguish between healthy and sick patterns and identify minute changes in thermal distributions.

Conventional methods for thermography-based breast cancer diagnosis relied on statistical descriptors, hand-crafted features, and traditional machine learning classifiers [8,9]. Although these techniques work well in certain situations, their sensitivity to patient variability and imaging circumstances limits their generalizability. By automatically deriving discriminative representations, convolutional neural networks (CNNs) have shown greater performance [10–12]. However, most CNN-based models for breast cancer thermography now in use either solely use spatial characteristics or heavily rely on backbone networks, disregarding complementing frequency-domain information. As a result of this, interpretability, generalization, and model efficiency may be restricted [13,14].

To address these challenges, this work proposes a multi-branch hybrid model that simultaneously leverages spatial-domain and frequency-domain information. Spatial features are extracted directly from segmented thermograms, while complementary frequency features are derived from their Power Spectral Density (PSD) transformations. A Shapley value-based Gated Fusion mechanism then adaptively integrates the two streams, ensuring that salient and informative cues are emphasized while redundant information is suppressed. Furthermore, the model employs a lightweight backbone (MobileNetV3Small), striking a balance between computational efficiency and classification accuracy, thereby making it suitable for deployment in resource-constrained environments.

In summary, this paper makes the following **contributions**:

- We introduce a multi-branch hybrid deep learning framework that jointly exploits spatial-domain thermographic information and frequency-domain cues from PSD transformations.
- A Shapley value-based Gated Fusion mechanism is employed to adaptively integrate complementary features across modalities, enhancing discriminative capability.
- By adopting a lightweight backbone, the model achieves high diagnostic accuracy on the DMR-IR dataset, while maintaining efficiency, enabling practical deployment for large-scale screening.

2 Related Work

Medical infrared thermography has gained significant attention as a diagnostic tool, primarily due to its non-contact, non-ionizing, and cost-effective characteristics. Initial research efforts in this field concentrated on manually designed feature extraction methods, thermal pattern symmetry evaluation, and probabilistic feature characterization [8,9]. Although these methodologies showed promise in laboratory settings, their dependence on hand-crafted feature engineering compromised their effectiveness across heterogeneous patient populations and variable acquisition protocols.

Subsequently, investigators turned to conventional machine learning paradigms to address these limitations. Textural feature extraction techniques, including Gray Level Co-occurrence Matrix (GLCM) analysis, paired with classification algorithms such as support vector machines (SVMs), produced modest performance enhancements [15,16]. Additionally, symmetry-based and energy-derived metrics from breast thermal patterns were utilized for tumor identification [8]. Nevertheless, these feature-engineering approaches frequently struggled with cross-dataset generalization and demonstrated insufficient flexibility in capturing intricate pathological manifestations.

The emergence of deep learning technologies transformed thermal imaging-based breast malignancy screening. CNNs exhibited remarkable capability in autonomously extracting meaningful patterns from thermographic data [14,17]. Research utilizing VGG-inspired architectures [17], ResNet frameworks [18], and dual-input CNN configurations [15,19] achieved notable diagnostic accuracy gains. Additionally, genetic optimization algorithms and composite deep learning models contributed to enhanced classification outcomes [20,21]. Garyali et al. [22]. proposed a CNN-based AI system for breast cancer screening using thermograms from public databases (DMR-IR and DIA), achieving 93% accuracy, 93% sensitivity, 91% specificity, and 95% precision on a test set of 960 images. Their model incorporates data augmentation, a VGG16-inspired architecture, and a novel post-processing formula to predict cancer risk levels (low, medium, high) based on multiple images per patient, positioning thermography as a viable adjunct for early detection in resource-limited settings. However, the majority of CNN-based methodologies focused solely on spatial feature extraction, potentially missing valuable spectral information inherent in thermal signal characteristics.

Contemporary research has started investigating multi-source and cross-modal data integration strategies. Sanchez-Cauce and colleagues [15] merged thermal imagery with patient metadata through unified CNN architectures, whereas Tsietso et al. [19] combined multiple thermal perspectives with physiological measurements. Similarly, Attallah [13] employed multi-pathway CNN feature extraction for premature breast cancer identification. While these methodologies enhanced model robustness, they frequently utilized computationally intensive backbone networks like Inception and ResNet, restricting their applicability in resource-limited clinical settings.

Our approach distinguishes itself by synthesizing spatial and spectral domain information through the integration of thermal images with their PSD analyses via a bifurcated path with CNN architectures. An adaptive Shapley value-based-Gated Fusion strategy dynamically balances the influence of each domain-specific pathway, promoting enhanced feature discrimination while preventing information redundancy. Additionally, by implementing a compact backbone network (MobileNetV3Small), our methodology achieves an optimal trade-off between diagnostic accuracy and computational demands, rendering it especially appropriate for practical deployment in clinical screening applications where system scalability and processing efficiency are paramount considerations.

3   Dataset

The thermal breast images used in this study come from the *Imagens e Matrizes da Tese de Thiago Alves Elias da Silva* dataset, part of the Database for Mastology Research with Infrared Image (DMR-IR). This dataset was compiled at the University Hospital of the Federal University of Fluminense, Niterói, Brazil, and contains thermographic examinations paired with clinical records. It includes data from 32 patients with breast abnormalities. Classification followed the Breast Imaging-Reporting and Data System (BI-RADS), where individuals labeled BI-RADS 1 or 2 are healthy controls, and those diagnosed with malignancies are cancer-positive. Documented abnormalities include fluidity, phyllodes tumors, and liquid nodules. Each thermal image was captured at a resolution of 640×480 pixels.

For this analysis, the images were organized into two categories: pathological cases (*Doentes*) and healthy cases (*Saudáveis*). Expert annotators segmented the breast region of interest (ROI) in each image to precisely localize diagnostically relevant regions. The segmented images were stored in dedicated *Segmentadas* folders according to their category.

Formally, the dataset is represented as

$$\mathcal{D} = \{(\mathbf{I}_i, y_i)\}_{i=1}^{N},$$

where $\mathbf{I}_i \in \mathbb{R}^{H \times W}$ denotes the i-th manually segmented grayscale thermal image, and $y_i \in \{0, 1\}$ is the binary label indicating cancer-positive (1) or healthy case (0).

An 80:20 stratified split was applied to partition the data into training and testing subsets, while maintaining the original proportion of positive and negative cases.

4   Proposed Methodology

The proposed framework for breast cancer classification is a multi-branch, hybrid deep learning model. It simultaneously processes spatial information from segmented images and frequency-domain information from their Power Spectral

Density (PSD) transformations. The features from these parallel streams are adaptively merged using a Shapley value-based Gated Fusion mechanism and combined with features from a pre-trained backbone. The complete methodology is detailed below.

4.1   Image Preprocessing and PSD Transformation

The preprocessing stage standardizes the input image and generates a corresponding frequency-domain representation for each image.

Let an input grayscale thermal image be denoted as $\mathbf{I}_i \in \mathbb{R}^{H \times W}$. A median blur filter with a kernel size $k = 3$ is first applied to reduce high-frequency noise:

$$\mathbf{I}'_i = \mathrm{MedianBlur}(\mathbf{I}_i, k = 3). \tag{1}$$

The blurred image $\mathbf{I}'_i$ is then resized to a fixed spatial resolution of 224×224 pixels and normalized to the range $[0, 1]$ to create the spatial input image $\mathbf{X}_{\mathrm{spatial},i}$:

$$\mathbf{X}_{\mathrm{spatial},i} = \frac{\mathrm{Resize}(\mathbf{I}'_i, (224, 224))}{255.0}. \tag{2}$$

To capture frequency-domain characteristics, the log Power Spectral Density (PSD) of the normalized image is computed. This involves applying a 2D Fast Fourier Transform ($\mathcal{F}$), shifting the zero-frequency component to the center, and computing the log-scaled power spectrum:

$$\mathbf{P}_i = \log(1 + |\mathcal{F}_{\mathrm{shift}}(\mathcal{F}(\mathbf{X}_{\mathrm{spatial},i}))|^2). \tag{3}$$

The resulting PSD image $\mathbf{P}_i$ is then normalized to the range $[0, 1]$ to create the frequency input image $\mathbf{X}_{\mathrm{freq},i}$.

Finally, both the single-channel spatial image $\mathbf{X}_{\mathrm{spatial},i}$ and the frequency image $\mathbf{X}_{\mathrm{freq},i}$ are converted to three-channel images by duplicating their respective channels, yielding the final model inputs $\mathbf{X}_{\mathrm{IR},i} \in \mathbb{R}^{224 \times 224 \times 3}$ and $\mathbf{X}_{\mathrm{PSD},i} \in \mathbb{R}^{224 \times 224 \times 3}$.

4.2   Network Architecture

The model architecture is composed of three parallel branches: a spatial CNN branch, a frequency CNN branch, and a pre-trained MobileNetV3Small backbone. Their outputs are combined for final classification (Fig. 1).

Spatial and Frequency CNN Branches. Two identical, parallel CNNs are employed to process thermograms ($\mathbf{X}_{\mathrm{IR},i}$) and PSD ($\mathbf{X}_{\mathrm{PSD},i}$) inputs, respectively. Each branch consists of three convolutional stages. A stage is composed of a 3×3 convolutional layer, followed by batch normalization, a rectified linear unit (ReLU) activation function, and a max-pooling layer for downsampling. The number of feature channels increases across the stages as $32 \rightarrow 64 \rightarrow 128$.

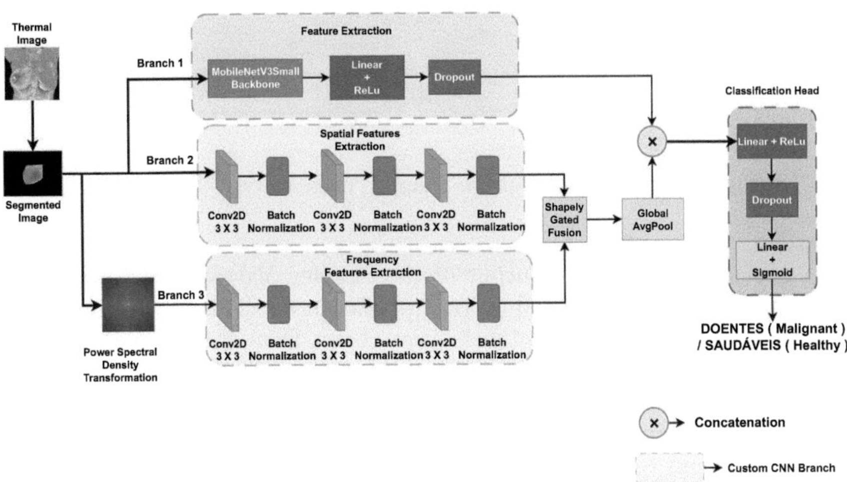

Fig. 1. The architecture represents the proposed architecture of the Shapley Value-based Gated Feature Fusion of Multi-branch Deep Learning Framework for Breast Cancer Screening of Thermal Images

Pre-Trained Backbone. To leverage features learned from a large-scale dataset, a pre-trained MobileNetV3Small model, with weights frozen from ImageNet, is used as a third feature extractor. It takes the thermogram input $\mathbf{X}_{\mathrm{IR},i}$, and its output is processed by a global average pooling layer, a dense layer with 128 units and ReLU activation, and a dropout layer with a rate of $p = 0.4$.

Shapley Value-Based Gated Fusion Mechanism. The feature maps from the spatial branch ($\mathbf{F}_{\mathrm{spatial}}$) and the frequency branch ($\mathbf{F}_{\mathrm{freq}}$), both in $\mathbb{R}^{H' \times W' \times 128}$, are adaptively combined using a Shapley Value-based Gated Fusion layer [23]. The working of the Shapley Value-based Gated Fusion Mechanism is shown in 2. This layer computes gating weights, $\mathbf{w}_{\mathrm{spatial}}$ and $\mathbf{w}_{\mathrm{freq}}$, which control the contribution of each branch:

$$\mathbf{u}_{\mathrm{spatial}} = \sigma(\mathrm{Conv}_{1 \times 1}(\mathbf{F}_{\mathrm{spatial}})) \tag{4}$$

$$\mathbf{u}_{\mathrm{freq}} = \sigma(\mathrm{Conv}_{1 \times 1}(\mathbf{F}_{\mathrm{freq}})) \tag{5}$$

$$\mathbf{w}_{\mathrm{spatial}} = \frac{\mathbf{u}_{\mathrm{spatial}}}{\mathbf{u}_{\mathrm{spatial}} + \mathbf{u}_{\mathrm{freq}} + \epsilon} \tag{6}$$

$$\mathbf{w}_{\mathrm{freq}} = \frac{\mathbf{u}_{\mathrm{freq}}}{\mathbf{u}_{\mathrm{spatial}} + \mathbf{u}_{\mathrm{freq}} + \epsilon} \tag{7}$$

where σ is the sigmoid function and ϵ is a small constant for numerical stability. The final fused feature map $\mathbf{F}_{\mathrm{fused}}$ is a weighted sum:

$$\mathbf{F}_{\mathrm{fused}} = \mathbf{w}_{\mathrm{spatial}} \odot \mathbf{F}_{\mathrm{spatial}} + \mathbf{w}_{\mathrm{freq}} \odot \mathbf{F}_{\mathrm{freq}}, \tag{8}$$

where $\odot$ denotes element-wise multiplication.

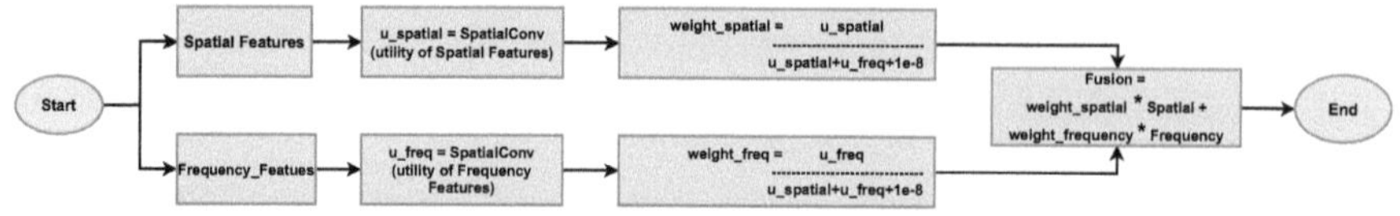

Fig. 2. An illustration of the Shapely-Value based Gated Fusion

Classification Head. The output of the Shapley Value-based Gated Fusion layer is passed through a global average pooling layer. This pooled feature vector is then concatenated with the feature vector from the MobileNetV3Small backbone. The resulting combined vector is fed into a final dense layer with 128 units (ReLU activation), followed by a dropout layer with probability $p = 0.5$. The classification head consists of a single neuron with a sigmoid activation function to output the predicted probability of malignancy $\hat{y} \in (0, 1)$.

5 Experimental Setup and Results

5.1 Experimentation Setup

The model was implemented in PyTorch within the Kaggle environment, utilizing an NVIDIA Tesla T4 GPU for all experiments. The network is trained using Adam optimizer with an initial learning rate $\eta = 10^{-3}$. The loss function is the binary cross-entropy, defined as:

$$\mathcal{L} = -\frac{1}{M} \sum_{i=1}^{M} \left[y_i \log \hat{y}_i + (1 - y_i) \log(1 - \hat{y}_i) \right], \tag{9}$$

where M is the batch size, $y_i \in \{0, 1\}$ is the ground-truth label, and $\hat{y}_i$ is the predicted probability.

Training is conducted for a maximum of 100 epochs with a batch size of $M = 16$. A set of callbacks is employed to ensure robust training: ModelCheckpoint saves the model with the best validation accuracy and ReduceLROnPlateau decreases the learning rate by a factor of 0.2 if the validation loss stagnates for 5 epochs. The model's performance is evaluated based on accuracy and the Area Under the Curve (AUC).

5.2 Comparison with Past Methods

The experimental evaluation on the DMR-IR dataset demonstrates the superior performance of the proposed model compared to existing methodologies. As presented in Table 1, our approach achieves exceptional results across all evaluation metrics, with perfect scores of 1.0000 for accuracy, precision, recall, and F1-score. This outstanding performance indicates the model's ability to achieve flawless classification without any misclassified samples. Among the competing

Table 1. Performance comparison on the DMR-IR dataset.

Work Ref.	Accuracy	Precision	Recall	F1-Score
Dharani et al. [16]	0.9680	–	–	–
Nogales et al. [20]	0.9385	0.9666	–	–
Alshehri et al. [24]	0.9980	0.9976	0.9985	0.9980
Khodadadi et al. [26]	0.9865	0.9878	0.9979	–
Shojaedini et al. [27]	0.9230	0.9300	–	–
Jayagayathri et al. [29]	0.9773	1.000	0.9835	0.9787
Lescarbeault et al. [30]	0.9620	0.9940	0.9500	0.9710
Tsietso et al. [19]	0.9048	0.9333	–	–
Garyali et al. [22]	0.93	0.95	0.93	0.94
Proposed Model	1.0000	1.0000	1.0000	1.0000

methods, several approaches demonstrate varying degrees of effectiveness. Dharani et al. [16] achieves an accuracy of 0.9680, though complete metric evaluation is not available. Nogales et al. [20] reports 0.9385 accuracy with 0.9666 precision, but recall and F1-score metrics are not provided. Alshehri et al. [24] presents highly competitive performance with comprehensive metrics: 0.9980 accuracy, 0.9976 precision, 0.9985 recall, and 0.9980 F1-score, representing the closest performance to our proposed framework among all evaluated methods. Gupta et al. [25] achieves 0.9573 accuracy, while Khodadadi et al. [26] demonstrates strong performance with 0.9865 accuracy, 0.9878 precision, and 0.9979 recall. Shojaedini et al. [27] reports 0.9230 accuracy with 0.9300 precision, and Tang et al. [28] achieves 0.9860 accuracy, 0.9737 precision, 0.9867 recall, and 0.9802 F1-score. Jayagayathri et al. [29] presents interesting results with 0.9773 accuracy, perfect precision (1.000), 0.9835 recall, and 0.9787 F1-score. Lescarbeault et al. [30] achieves 0.9620 accuracy, 0.9940 precision, 0.9500 recall, and 0.9710 F1-score, while Tsietso et al. [19] reports 0.9048 accuracy with 0.9333 precision.

It is worth noting that several studies report incomplete metric evaluations, providing only accuracy or limited performance measures, which restricts comprehensive comparative analysis. Despite the strong performance demonstrated by various competing methods, particularly Alshehri et al. [24], our proposed model's achievement of perfect scores across all metrics establishes a new benchmark for the DMR-IR dataset, demonstrating its robustness and reliability for practical applications.

5.3   GradCam

We have used Gradient-weighted Class Activation Mapping (Grad-CAM) to better understand the input properties that impact the model in classification. The Fig. 3 shows both the input image and the associated Grad-CAM visualization. This visualization has a heatmap that highlights sections of the image, where the

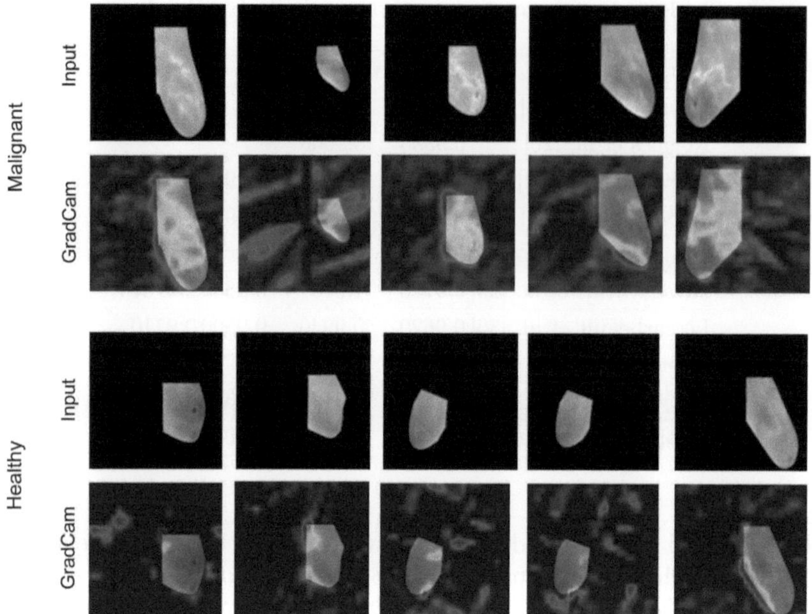

Fig. 3. The following figure represents the GradCam Images of 8 different segmented samples of DMR-IR dataset

model concentrates its attention, showing the presence or absence of Malignant images and directing its decision-making process. In malignant images, the model focuses attention on specific areas, but in healthy images, it spreads attention across the cell. This highlights the model's attention capacity to make correct classifications depending on the presence or absence of malignant images.

5.4   Ablation Study

To validate the effectiveness of our proposed model, we conducted an extensive ablation study using different backbone architectures, feature combinations, and fusion strategies. The results are summarized in Table 3.

We first evaluate traditional backbone networks such as EfficientNetB0/B4, ResNet50, AlexNet, and MobileNet with psd+spatial features under a Shapley-Gated fusion scheme. While these models achieved strong performance (accuracy ranging from 98.83% to 99.61% and AUC up to 99.99%), they generally required a larger number of parameters (up to 85.17 MB in the case of AlexNet).

Next, we analyzed the contribution of different feature modalities using MobileNetV3Small. When using only spatial or only psd features, the performance dropped noticeably (accuracy 98.31% and 99.22%, respectively), highlighting the necessity of combining both feature domains for robust representation.

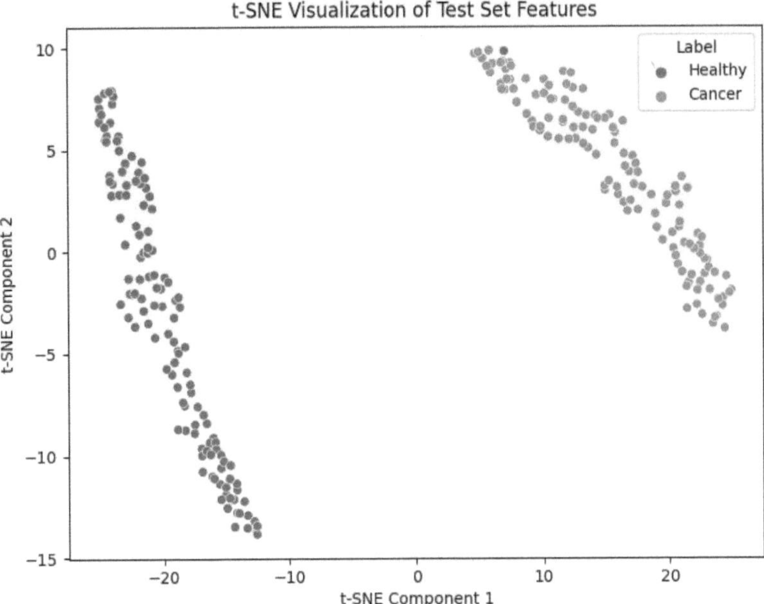

Fig. 4. t-SNE visualization of feature distribution for thermal breast images showing class separability between cancerous and healthy samples

We further examined different feature fusion strategies. Concatenation and addition fusion achieved promising performance (accuracy 99.61% and 99.92%), but they either slightly reduced the discriminative capability or increased parameter redundancy.

Finally, our proposed model (MobileNetV3Small with PSD+spatial features and Shapley Values-based Gated fusion) achieved the best overall results, obtaining 100% accuracy and 100% AUC, while maintaining the lowest parameter count (282,177; 1.08 MB) among all multi-feature fusion approaches. This demonstrates that the Shapley-Gated fusion strategy effectively leverages complementary information from both feature domains while preserving model compactness, making it both accurate and lightweight.

5.5   Cross-Validation and T-SNE Visualization

To ensure the robustness and generalizability of the proposed two-class classification model, a **5-fold cross-validation** strategy was employed. The dataset was randomly divided into five equal folds, where each fold was used once as the validation set while the remaining four folds served as the training set. Table 2 presents the cross-validation results in terms of Accuracy (Acc), Precision, Recall, and F1-score. The model consistently achieved near-perfect performance across all folds, with three folds achieving **100%** in every metric, while the remaining folds still maintained $\geq 98.99\%$ accuracy with balanced precision and recall. Such minimal inter-fold variance demonstrates the model's stability and its strong ability to avoid overfitting.

Table 2. Five-fold cross-validation results for the proposed model.

Fold No.	Acc (%)	Precision (%)	F1 (%)	Recall (%)
1	100.00	100.00	100.00	100.00
2	100.00	100.00	100.00	100.00
3	99.99	99.99	99.99	99.99
4	98.99	98.98	98.99	98.99
5	100.00	100.00	100.00	100.00
Mean	99.79	99.79	99.80	99.80

To further evaluate the learned feature representations, we applied **t-Distributed Stochastic Neighbor Embedding (t-SNE)** on the high-dimensional features extracted from the penultimate layer of the network. The resulting two-dimensional embedding revealed **two well-separated clusters**, each corresponding to one of the target classes shown in Fig. 4. The distinctive inter-cluster distance confirms that the network captures highly discriminative features, providing a strong qualitative explanation for the excellent cross-validation performance.

Table 3. Ablation results of different backbones, feature types, and fusion strategies. Our proposed model (last row) achieves the best performance with the lowest parameter count. Accuracy and AUC scores are in %.

Backbone	Features	Fusion	Accuracy	AUC	Parameters
EfficientNetB0	PSD+Spatial	Gated	98.83	99.81	417,409 (1.59 MB)
EfficientNetB4	PSD+Spatial	Gated	98.83	99.99	482,945 (1.84 MB)
ResNet50	PSD+Spatial	Gated	99.22	99.97	515,713 (1.97 MB)
AlexNet	PSD+Spatial	Gated	99.61	99.98	22,326,017 (85.17 MB)
MobileNet	PSD+Spatial	Gated	100	100	384,641 (1.47 MB)
MobileNetV3Small	Spatial	None	98.31	99.20	294,273 (1.12 MB)
MobileNetV3Small	PSD	None	99.22	99.80	294,273 (1.12 MB)
MobileNetV3Small	PSD+Spatial	Concatenate	99.61	99.95	310,657 (1.19 MB)
MobileNetV3Small	PSD+Spatial	Addition	99.92	99.98	294,273 (1.12 MB)
MobileNetV3Small	**PSD+Spatial**	**Gated**	**100**	**100**	**282,177 (1.08 MB)**

6 Conclusion

This study presents a novel multi-branch hybrid deep learning framework for breast cancer classification using thermal infrared images. By jointly exploiting spatial-domain representations and frequency-domain features via Power Spectral Density (PSD) transformation, the model leverages complementary thermal patterns. The adaptive Shapley Value-based Gated Fusion mechanism effectively integrates these feature streams, while a lightweight MobileNetV3Small backbone balances accuracy and computational efficiency. Experimental results demonstrate that the proposed architecture outperforms existing state-of-the-art methods, achieving high accuracy and robust generalization on the publicly available DMR-IR dataset.

Despite these promising results, several limitations require consideration in the future. The dataset size is relatively small, potentially limiting the generalizability of the model to broader populations. The model's performance on thermograms with varying acquisition protocols and from diverse demographic groups remains to be evaluated. Future work will focus on expanding dataset diversity, automating ROI segmentation, and exploring temporal dynamics using thermal image sequences to further improve diagnostic performance and robustness.

References

1. World Health Organization. Breast cancer: Prevention and control. WHO Fact Sheet (2021)
2. Siegel, R.L., Miller, K.D., Wagle, N.S., Jemal, A.: Cancer statistics CA: A Can. J. Clin. **70**(1), 7–30 (2020)
3. Yaffe, M.J.: Breast density and cancer risk: the potential impact of digital mammography and tomosynthesis. Breast Cancer Res. **13**(2), 123–134 (2011)
4. Boyd, N.F., Yaffe, M.J.: Mammographic density and breast cancer risk: current understanding and future prospects. Breast Cancer Res. **16**(1), 212 (2014)
5. Elmore, J.G., Fletcher, S.W., Kramer, B.S.: Screening for breast cancer. JAMA **293**(10), 1245–1256 (2005)
6. Head, J.F., Elliott, R.L.: Infrared imaging: making progress in fulfilling its medical promise. IEEE Eng. Med. Biol. Mag. **18**(3), 52–54 (1999)
7. Ng, E.Y.K.: A review of thermography as promising non-invasive detection modality for breast tumor. Int. J. Therm. Sci. **48**(5), 849–859 (2009)
8. Sathish, T., Subashini, R.: Detection of breast cancer using ir thermograms and statistical features. In: 2018 IEEE International Conference on Communication and Signal Processing (ICCSP), pp. 1105–1109 (2018)
9. Marcus, C.L., Flower, W.B.: Analysis of breast thermograms using statistical and graphical methods. Quant. InfraRed Thermography J. **3**(2), 87–101 (2006)
10. Krizhevsky, A., Sutskever, I., Hinton, G.E.: Imagenet classification with deep convolutional neural networks. Adv. Neural Inf. Process. Syst. (NeurIPS) **25**, 1097–1105 (2012)
11. Simonyan, K., Zisserman, A.: Very deep convolutional networks for large-scale image recognition. In: ICLR (2015)

12. He, K., Zhang, X., Ren, S., Sun, J.: Deep residual learning for image recognition. In: IEEE Conference on Computer Vision and Pattern Recognition (CVPR), pp. 770–778 (2016)
13. Attallah, O.: Harnessing infrared thermography and multi-convolutional neural networks for early breast cancer detection. Sci. Rep. **15**(1), 27464 (2025)
14. Torres-Galván, J.C., et al.: Deep convolutional neural networks for classifying breast cancer using infrared thermography. Quant.3 InfraRed Thermography J. **19**, 283–294 (2022)
15. Sánchez-Cauce, R., Pérez-Martín, J., Luque, M.: Multi-input convolutional neural network for breast cancer detection using thermal images and clinical data. Comput. Methods Programs Biomed. **204**, 106045 (2021)
16. Dharani, N.P., Govardhini Immadi, I., Venkata Narayana, M.: Enhanced deep learning model for diagnosing breast cancer using thermal images. Soft Comput. **28**(13), 8423–8434 (2024)
17. Ahmed, F., et al.: Breast cancer detection with vgg16: a deep learning approach with thermographic imaging. Breast Cancer Detection with VGG16 (2019)
18. Mirasbekov, Y., et al.: Fully interpretable deep learning model using ir thermal images for possible breast cancer cases. Biomimetics **9**(12), 609 (2024)
19. Tsietso, D., et al.: Multi-input deep learning approach for breast cancer screening using thermal infrared imaging and clinical data. IEEE Access **11**, 52101–52116 (2023)
20. Nogales, A., Perez-Lara, F., García-Tejedor, Á.J.: Enhancing breast cancer diagnosis with deep learning and evolutionary algorithms: a comparison of approaches using different thermographic imaging treatments. Multimedia Tools Appl. **83**(14), 42955–42971 (2024)
21. Vijaya Madhavi and Christy Bobby Thomas: Multi-view breast thermogram analysis by fusing texture features. Quant. InfraRed Thermography J. **16**(1), 111–128 (2019)
22. Garyali, P., Ranjbar, I., Movahedi, S.: A novel thermography-based artificial intelligence-powered solution for screening. In: Artificial Intelligence over Infrared Images for Medical Applications and Medical Image Assisted Biomarker Discovery: First MICCAI Workshop, AIIIMA 2022, and First MICCAI Workshop, MIABID 2022, Held in Conjunction with MICCAI 2022, Singapore, September 18 and 22, 2022, Proceedings, vol. 13602, p. 34. Springer (2022)
23. Roth, A.E.: Introduction to the shapley value. The Shapley value 1 (1988)
24. Alshehri, A., AlSaeed, D.: Breast cancer diagnosis in thermography using pre-trained vgg16 with deep attention mechanisms. Symmetry **15**(3), 582 (2023)
25. Gupta, T., Agrawal, R.K., Sangal, R., Rao, S.A.: Performance evaluation of thermography-based computer-aided diagnostic systems for detecting breast cancer: an empirical study. ACM Trans. Comput. Healthcare **5**(4), 1–30 (2024)
26. Khodadadi, H., Nazem, S.: Improving cancer detection through computer-aided diagnosis: a comprehensive analysis of nonlinear and texture features in breast thermograms. PLoS ONE **20**(5), e0322934 (2025)
27. Seyed Vahab Shojaedini and Bahram Bahramzadeh: A new method for promoting the detection of breast cancer in thermograms: applying deep autoencoders for eliminating redundancies in parallel with preserving independent components. J. Ambient. Intell. Humaniz. Comput. **15**(12), 4085–4099 (2024)
28. Tang, Y., Zhou, D., Flesch, R.C., Jin, T.: A multi-input lightweight convolutional neural network for breast cancer detection considering infrared thermography. Expert Syst. Appl. **263**, 125738 (2025)

29. Jayagayathri, I., Mythili, C.: Ob-tsasa guided soft voting ensemble classifier for breast cancer classification using thermal infrared images. IETE J. Res. **71**(2), 710–728 (2025)
30. Lescarbeault, É., Seoud, L.: Breast cancer detection from thermal images using asymmetries in learned texture vectors. In: 2025 IEEE 22nd International Symposium on Biomedical Imaging (ISBI), pp. 1–5. IEEE (2025)

An End-to-End GAN-CNN Framework for Early Breast Anomaly Detection Using Thermal Imaging

Hector Espinos-Morato[1], Carlos Barroso[2]([✉]), Johanna Angulo[1],
Alex De María Espiritusanto[1], Jenny Paula Aguilar Guiñez[1],
and Cesáreo Tello Sánchez[1]

[1] Universidad Europea de Valencia (UEV), School of Sciences, Engineering and
Design, Valencia, Spain
`hector.espinos@universidadeuropea.es`,
`{22186166,224F2279,224J5856,224G8714}@live.uem.es`
[2] Universidad Europea de Madrid (UEM), Madrid, Spain
`22014885@live.uem.es`

Abstract. Infrared thermography is a low-cost, non-invasive technique with potential for early breast cancer detection. However, clinical adoption is hindered by limited data availability and high interpatient variability. This work proposes a novel end-to-end framework that combines Generative Adversarial Networks (GANs) and Convolutional Neural Networks (CNNs) to generate synthetic thermal images, perform automated breast segmentation, and classify anomalous patterns.

The Visual Lab Breast Thermography dataset was used with thermal normalization and automatic detection of the region of interest (ROI) based on morphological and thermal features. A CycleGAN architecture is trained to generate realistic images, evaluated using the Fréchet Inception Distance (FID) and Kernel Inception Distance (KID). Synthetic data were combined with real images to train the EfficientNet-B0 and ResNet50 classifiers (in permissive and strict configurations), achieving an accuracy of over 97% and an area under the ROC curve (AUC) of 0.977.

The results demonstrate the feasibility of integrating GAN-based augmentation with CNN classification for robust automated breast screening. This method offers a promising diagnostic support system, especially suited for low-resource or mobile healthcare environments.

Keywords: Breast thermography · Generative adversarial networks · Automatic segmentation · CNN classification · Fréchet inception distance · Data augmentation · Thermal imaging · Breast cancer detection · Deep learning

1 Introduction

Breast cancer remains one of the leading causes of mortality among women worldwide, with more than 2.3 million new cases reported in 2022 according to

GLOBOCAN [1]. Early detection plays a crucial role in improving the prognosis, survival rates, and patient quality of life [2]. Although mammography continues to be the standard screening method, it has notable limitations: exposure to ionizing radiation, reduced accuracy in dense breast tissue, and limited accessibility in low-resource settings [3–5].

In this context, infrared thermography emerges as a promising complementary technique [6,7]. It is non-invasive, radiation-free, low-cost, and suitable for mobile health environments [8]. Its ability to detect thermal abnormalities associated with tumor-related processes, such as angiogenesis, hypermetabolism, and inflammation, makes it a valuable candidate for early detection [9,10]. However, effective clinical application is hindered by interpatient thermal variability, lack of standardized datasets, and challenges in precise anatomical segmentation [11–14].

To address these issues, a comprehensive diagnostic system has been proposed that combines synthetic image generation using Generative Adversarial Networks (GANs) [15,16] with automatic classification through Convolutional Neural Networks (CNNs) [17,18]. Unlike traditional approaches that use GANs only for data augmentation [19,20], our framework integrates both GANs and CNNs into a unified diagnostic (see Fig. 1). This end-to-end architecture improves classifier performance and enhances robustness when training with small datasets [21].

2   Dataset and Preprocessing

2.1   Dataset Description

For system development and validation, we used the public *Visual Lab Breast Thermography* dataset from the Visual Lab research group at Universidade Federal Fluminense (Brazil), accessible through the portal http://visual.ic.uff.br/dmi. This dataset contains 280 frontal thermal breast images in matrix (`.txt`) format, acquired under standardized clinical conditions with associated clinical annotations [11]. It includes 179 images from patients with no signs of breast anomalies and 101 from patients with confirmed benign or malignant lesions.

The dataset was selected due to its public availability, frequent use in prior research, and harmonized acquisition protocol, which facilitates reproducibility and comparability [22]. Additionally, it provides relevant clinical metadata to support in-depth analysis.

2.2   Thermal Normalization and Validation

Each image is represented as a floating-point matrix of temperature values. To mitigate inter-patient variability, we applied *min-max normalization* per patient, scaling all values to the $[0, 1]$ range based on each image's individual minimum and maximum temperature [12]. This approach preserves the relative thermal structure across regions regardless of absolute temperature differences. Normalization was validated by numerical inspection of histograms to ensure anatomical coherence and symmetrical distribution.

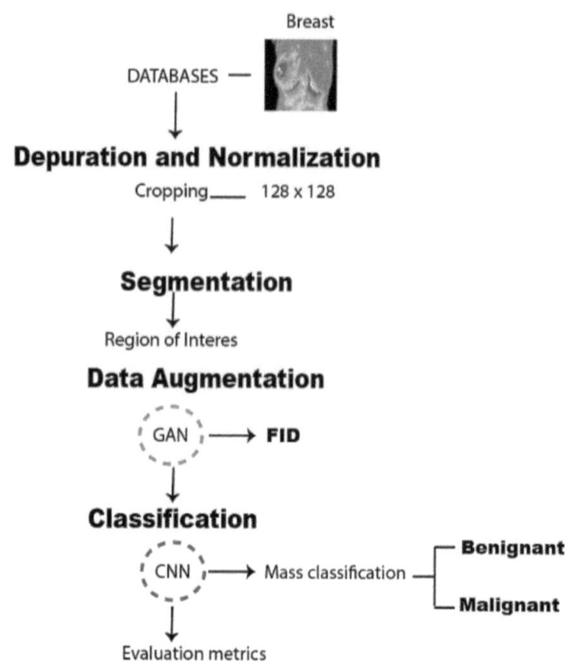

Fig. 1. Methodological framework of the project: synthetic image generation and automatic classification.

2.3   Automatic Segmentation of Regions of Interest

We developed a fully automated segmentation algorithm to isolate *breast anatomical regions* [23,24]. The region of interest (ROI) is determined based on five hierarchical criteria:

1. **Maximum thermal gradient:** highlights zones with abrupt temperature transitions.
2. **Morphological circularity:** favors rounded structures compatible with breast anatomy.
3. **Inferior anatomical location:** prioritizes lower sections of the image.
4. **Bilateral symmetry:** detects horizontal correspondence between breasts.
5. **Thermal artifact removal:** filters out rectangular edges and extreme noise.

This process standardizes all segmented images to a final resolution of 128 × 128 pixels, enabling consistent input to generative and classification models [25]. Segmentation quality was validated via inferior anatomical and morphological location inspection and bilateral thermal consistency checks. Figure 2 shows the logic sequence of the algorithm.

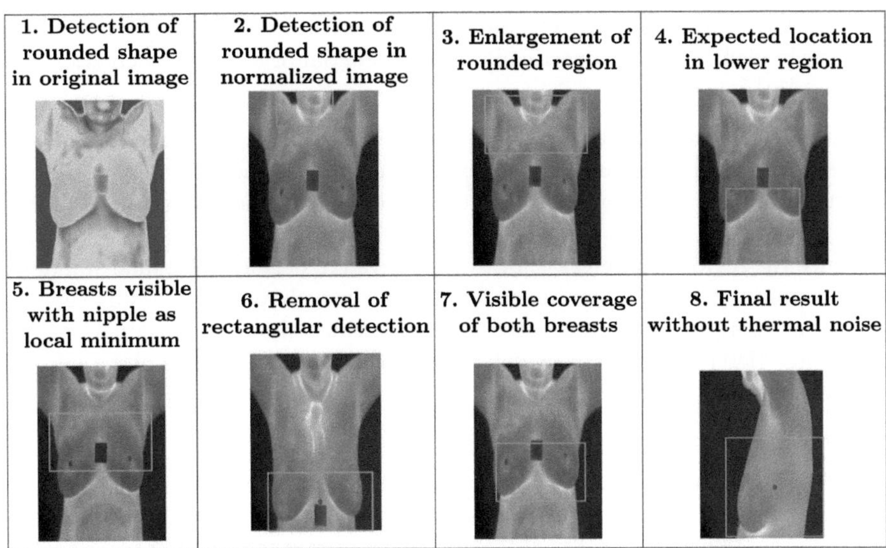

Fig. 2. Evolution of automatic thermal-image cropping criteria: from the initial detection of rounded regions, through the exclusion of thermal artefacts, to the final extraction of clean, coherent breast areas.

Detection of Rounded Shape in Original Image. The algorithm begins by identifying rounded contours in the original thermographic image, aiming to locate circular or elliptical regions that could correspond to breast outlines [23].

Detection of Rounded Shape in Normalized Image. The image is normalized to enhance contrast and improve robustness [12]. The shape detection step is repeated to confirm the presence and symmetry of rounded regions under uniform conditions.

Enlargement of the Rounded Region. The detected region is expanded to ensure full anatomical coverage, including peripheral tissue that may fall outside the initial detection zone [24].

Expected Location in Lower Region. A spatial constraint is applied: only rounded regions located in the lower half of the image are retained, as this is the typical anatomical position of the breasts in frontal thermography [11].

Breasts Visible with Nipple as Local Minimum. The thermal profile is analyzed to detect a local temperature minimum corresponding to the nipple area, reinforcing the anatomical plausibility of the detection [25].

Removal of Rectangular Detection. False positives—especially those shaped as rectangular thermal artefacts—are discarded using geometric filtering and edge consistency rules [23].

Visible Coverage of both Breasts. The algorithm verifies that both breasts are clearly visible and symmetrically framed. Partial or one-sided detections are rejected [11].

Final Result without Thermal noise. Morphological operations and intensity-based filtering are applied to remove residual thermal noise and background artefacts, resulting in a clean and anatomically coherent region of interest (ROI) [24].

Once the region of interest is located, the system performs the final standardized cropping, generating an image focused exclusively on the relevant anatomical area. This process is illustrated in Fig. 3, which shows the original normalized image (left), the automatic detection of the breast region (center), and the resulting cropped image (right).

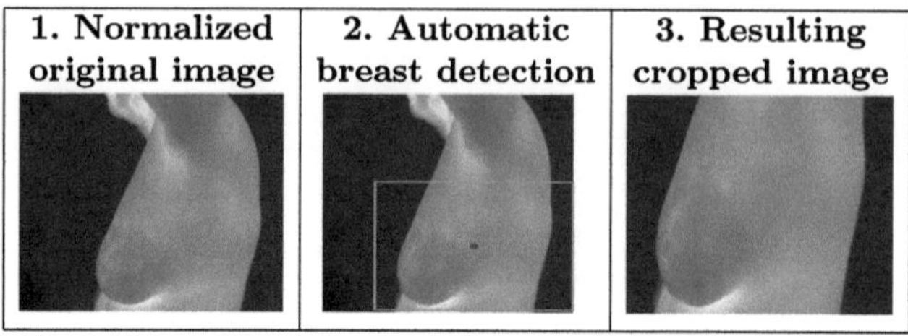

Fig. 3. Standardization of images in the automatic cropping process: normalized image, automated breast detection, and final crop with high clinical relevance.

3   Methodology

The proposed diagnostic framework is structured into two core components: (1) synthetic thermal image generation using a custom-designed GAN and (2) binary classification using deep convolutional neural networks [16, 26]. Both components are optimized for scenarios with limited datasets and computing resources.

3.1   Custom GAN Architecture

The Generative Adversarial Network (GAN) architecture employed in this study incorporates the CycleGAN model, which is well-suited for unpaired image-to-image translation tasks [27]. This choice is particularly appropriate given the limited availability and lack of paired datasets in the domain of thermal breast imaging [7].

CycleGAN's core mechanism of cycle consistency enforces that an image translated from the source domain (real thermal images) to the target domain (synthetic thermal images) and back remains faithful to the original [27]. This approach robustly preserves critical anatomical and thermal features critical for clinical validity, maintaining breast contours and thermographic gradients under unpaired conditions and minimizing hallucinations consistent with the reported FID/KID and downstream classification gains. Quantitative evaluation using established image quality metrics further supports the robustness of the architecture [28,29]. Specifically, the model achieved a Fréchet Inception Distance (FID) score of 38.4 and a Kernel Inception Distance (KID) of 0.021, values indicative of high-quality and realistic synthetic images within this application domain. Visual inspections corroborate these findings, showing morphologically coherent and thermally consistent images indistinguishable from real samples to expert observers.

Although simpler GAN variants with basic convolutional layers and no cycle consistency or conditioning mechanisms might struggle to capture the complex patterns inherent in medical thermography [16], the combination of CycleGAN architecture and rigorous evaluation protocols employed here provides sufficient robustness. This supports the generation of synthetic thermal images that effectively enrich training datasets, enhancing subsequent classification performance [21].

In summary, while there is always room for experimentation with newer or more complex GAN variants (e.g., SPADE, StyleGAN2), the current CycleGAN-based approach represents a robust and validated solution for realistic thermal image synthesis in breast cancer detection workflows [30].

Synthetic-To-Real Ratio. To clarify the contribution of synthetic images, we systematically varied the synthetic-to-real ratio in the training dataset. In the main configuration, the number of synthetic images matched the number of real samples (1:1 ratio), increasing the dataset from 280 to 560 images. Additional experiments with 0.5:1, 2:1, and 4:1 ratios confirmed that performance improvements plateau beyond a 2:1 ratio.

3.2 CNN Architectures for Classification

For the task of thermal image classification, two well-established convolutional neural network (CNN) architectures were selected: EfficientNetB0 and ResNet50 [31,32]. Both models were trained from scratch, eschewing transfer learning, due to the specific thermal characteristics of the dataset which differ substantially from typical RGB image domains [33]. Two distinct training configurations were devised to explore the trade-off between regularisation and learning speed:

- **Strict configuration:** Employing a dropout rate of 50% and a learning rate of 0.0005. This setup aims to enforce strong regularization to mitigate overfitting risks inherent to limited datasets [20].
- **Permissive configuration:** Utilising a lower dropout rate of 20% and a higher learning rate of 0.0015. This allows faster model adaptation to the training data but carries an increased risk of overfitting.

All architectures were optimized using the Adam optimizer, with binary cross-entropy as the loss function [17]. Early stopping and stratified 5-fold cross-validation were employed to ensure robust model generalization [22]. We explicitly ensured patient-level stratified 5-fold CV (no subject appears in both training and test). CycleGAN was trained only on the training folds within each split, and synthetic images were generated exclusively from training data to augment learning, while test folds remained strictly unseen Performance was evaluated using a comprehensive suite of metrics including accuracy, precision, recall, F1-score, and area under the ROC curve (AUC-ROC) [34].

Comparative experiments assessed model performance on original real thermal images versus an augmented dataset enriched with GAN-generated synthetic samples [19]. This two-stream pipeline—comprising GAN-based data augmentation followed by CNN classification—demonstrated improved robustness and generalization capabilities, particularly in low-resource scenarios [10].

EfficientNetB0 Architecture

Stem: Initial convolutional layer with 32 filters of size 3×3 and stride 2.
Mobile Inverted Bottleneck Convolution (MBConv) Blocks: Core of the network, performing inverted residuals with expansion, depthwise convolutions (3×3 or 5×5), and squeeze-and-excitation (SE) blocks to recalibrate channel-wise features. (e denotes the expansion factor for bottleneck layers).
Head: Final convolution, global average pooling, and fully connected classification layer.

The layer-wise summary of the EfficientNetB0 architecture can be seen in Table 1.

Table 1. EfficientNetB0 Layer-wise summary

Stage	Operator	Resolution	Channels	Layers
1	Conv 3×3	224×224	32	1
2	MBConv (k3 $\times$ 3, e1)	112×112	16	1
3	MBConv (k3 $\times$ 3, e6)	112×112	24	2
4	MBConv (k5 $\times$ 5, e6)	56×56	40	2
5	MBConv (k3 $\times$ 3, e6)	28×28	80	3
6	MBConv (k5 $\times$ 5, e6)	14×14	112	3
7	MBConv (k5 $\times$ 5, e6)	14×14	192	4
8	MBConv (k3 $\times$ 3, e6)	7×7	320	1
9	Conv 1×1 + Pool + FC	7×7	1280	1

ResNet50 Architecture

Initial Conv: 7×7 convolution with 64 filters and stride 2, followed by max pooling.
Residual Blocks: Four stages of bottleneck residual blocks composed of three convolutions (1×1, 3×3, 1×1) with skip connections. Each stage reduces spatial resolution and increases channels. *Bottleneck residual blocks* consist of 1×1 convolutional layers for dimensionality reduction and restoration with the 3×3 convolution in between, facilitating deeper architectures.
Final Layers: Global average pooling followed by a fully connected layer for classification.

The layer-wise summary of ResNet50 architecture it can see in Table 2.

Table 2. ResNet50 Layer-wise summary

Stage	Operator	Resolution	Channels	Layers
1	Conv7×7 + MaxPool	112×112	64	1
2	Conv Bottleneck Blocks	56×56	256	3
3	Conv Bottleneck Blocks	28×28	512	4
4	Conv Bottleneck Blocks	14×14	1024	6
5	Conv Bottleneck Blocks	7×7	2048	3
6	Global Average Pooling + FC	1×1	1000	1

Architectural Advantages and Limitations

EfficientNetB0. is characterized by its compound scaling method, balancing network depth, width, and resolution to optimize accuracy and efficiency [31]. It leverages depthwise separable convolutions to reduce parameter count and computational cost, making it highly efficient for deployment on resource-constrained environments [31]. Its architectural design enables superior parameter utilization, resulting in competitive accuracy with significantly fewer parameters compared to larger models [35]. For edge-device feasibility, we report model size and compute, EfficientNet-B0 ($\approx$5.3 M parameters; $\approx$0.39 GFLOPs) and ResNet-50 ($\approx$25.6 M; $\approx$4.1 GFLOPs). However, due to its relatively modest capacity, it might exhibit limitations in representing highly complex thermal patterns if not trained appropriately [33].

ResNet50. employs residual learning through skip connections which effectively mitigate vanishing gradient problems, allowing much deeper networks to be trained [32]. This architecture excels at capturing diverse and complex features, often converging faster and achieving state-of-the-art accuracy in image classification tasks [36]. Its greater depth and representational power provide an advantage in modeling intricate thermal imaging patterns [37]. The trade-off is increased computational demand and a higher number of parameters, which can impact inference speed and resource consumption.

In conclusion, the choice of these architectures balances efficiency and expressive power, providing a robust foundation for thermal image classification. The empirical findings indicate that integrating synthetic data augmentation via GANs significantly enhances classification performance, enabling reliable and scalable thermal anomaly detection [21].

All code developed within the scope of this study is thoroughly documented, well-structured, and publicly available through a GitHub repository. This repository includes scripts for thermal image preprocessing, synthetic image generation models using GAN architectures (CycleGAN, SPADE, StyleGAN2, among others), as well as classifiers built with CNNs. It also contains utilities for automatic segmentation, metric evaluation (FID, KID, AUC-ROC), and results visualization. Publishing the code ensures traceability and reproducibility of the methodological process, in alignment with open science principles. The repository is accessible at: https://github.com/Alexdemaes1/Thermal_Images.

3.3   Execution Environment

All experiments were conducted on the LORCA (Laboratorio de Optimización de Recursos Computacionales Avanzados), that is a supercomputer, belonging to the Universidad Europea de Madrid. This high-performance computing environment is characterized by providing access to a powerful architecture. Specifically, training sessions were run on a node with the following specifications: 64 logical cores and 125.6 GB of RAM.

Access to the environment was performed via SSH tunnel, using the Bitvise tool and a port forwarded to localhost, allowing usage of Jupyter Notebook through the personal computer's browser. Although functional, this connection system introduced several relevant technical limitations. One of the most problematic was cache memory management during training. Generally, each complete experiment—a single CNN training run—lasted between 6 and 8 h, justifying the limited number of configurations explored and the need for careful planning in each execution.

4   Results

This section presents the experimental outcomes obtained during the generation of synthetic images and the classification of thermal breast images. We report both qualitative and quantitative results for each phase of the system pipeline.

4.1   Qualitative Evaluation of GAN Output

The GAN-generated images were visually assessed at multiple checkpoints during training. Figure 4 illustrates the progressive improvement in thermal consistency and anatomical structure across training epochs (Epochs 10, 50, 100,

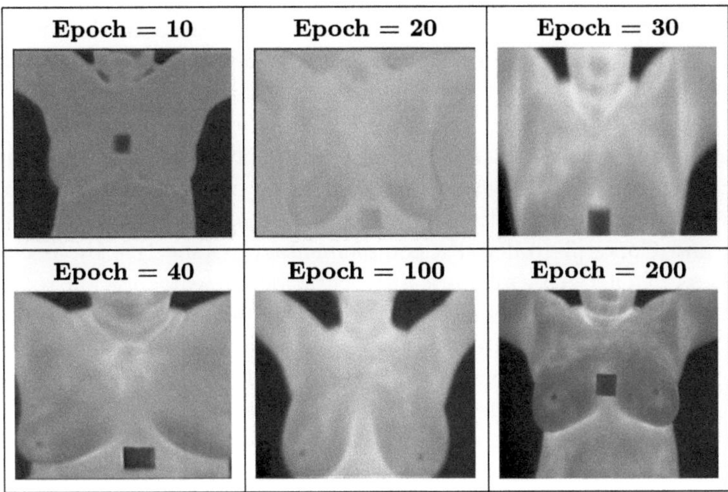

Fig. 4. Evolution of the GAN results in different *Epochs* in the *CycleGAN* model with Python.

and 200). Early-stage outputs displayed significant noise and deformation, while later epochs produced clearer and more anatomically coherent images.

Quantitative evaluation of GAN output was conducted using the Fréchet inception distance (FID) and the kernel inception distance (KID) [28, 29]. The final models achieved: FID: 38.4, KID: 0.08. These results indicate a high perceptual similarity between real and synthetic thermal images, validating the usefulness of the GAN for the enhancement of the dataset.

CNNs trained solely on real thermal images showed limited performance due to the small size of the dataset and the class imbalance. Table 3 summarizes the results (Table 4).

Table 3. Classification performance without synthetic data.

Model	Accuracy	Precision	Recall	F1-score	AUC
EfficientNetB0 (strict)	59.0%	67.0%	25.0%	0.54	0.702
ResNet50 (strict)	50.0%	50.0%	0.0%	0.00	0.605

Figure 5 presents the ROC curves for all models. The area under the curve (AUC) improved from near-random levels in baseline models to clinically acceptable values above 0.95 when synthetic data was used.

5   Discussion and Comparison

The integration of Generative Adversarial Networks (GANs) into breast thermography has enabled significant advances in image synthesis, data augmentation, and diagnostic support. Noteworthy contributions in this field include the work by Govindaraju and Kakileti [38], who developed a StyleGAN2-based model to generate high-fidelity synthetic breast thermal images. Their approach incorporated both pixel-level and frequency-based loss functions, resulting in synthetic images with remarkable visual and structural realism. Although their study did not include a classification pipeline, it provides a strong foundation for future diagnostic applications based on realistic thermal data synthesis.

Silva and Conci [39] offered a complementary perspective by assessing the effectiveness of DCGANs for data augmentation in segmentation tasks. Their results highlight the conditional benefits of synthetic data, which depend largely on how these images are integrated and validated within the target application. This reinforces the importance of end-to-end system design and thorough evaluation when employing generative models in medical imaging.

The study presented here proposes a unified GAN-CNN pipeline that extends these previous contributions by incorporating synthetic images directly into the classification process. A lightweight CycleGAN architecture was used to generate realistic thermal images from unpaired domains, and these images were employed to augment the training set for deep convolutional classifiers. As a result, the classification accuracy improved significantly, reaching 97.3% with an AUC of 0.977. This performance suggests that combining image synthesis

Table 4. Classification performance with synthetic image augmentation.

Model	Accuracy	Precision	Recall	F1-score	AUC
EfficientNetB0 (strict) + GAN	96.9%	96.3%	97.1%	0.96	0.977
ResNet50 (strict) + GAN	95.8%	95.2%	95.6%	0.95	0.973

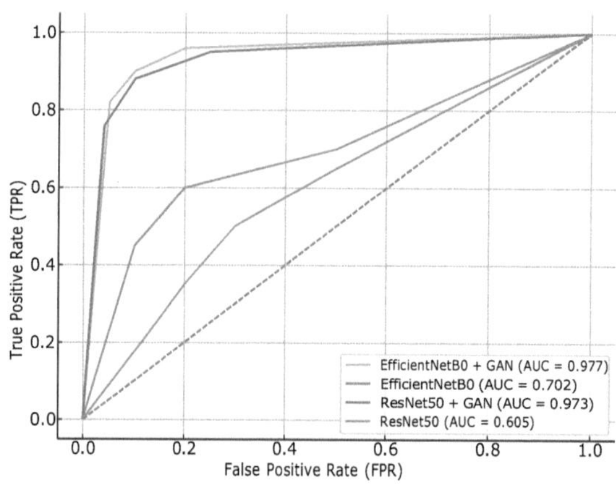

Fig. 5. Comparative ROC curves for EfficientNetB0 and ResNet50, with and without GAN-based augmentation.

with classification in a unified pipeline can produce synergistic improvements in diagnostic accuracy and model robustness.

Importantly, the methodological advances described in the aforementioned works are compatible with the approach taken in the present study. For instance, the advanced loss functions and high-resolution architectures introduced by Govindaraju and Kakileti [38] could be incorporated into future versions of the system to further enhance image quality. Similarly, the evaluative rigor advocated by Silva and Conci [39] aligns with the multi-metric validation framework adopted in this study, which includes FID, KID, and ROC-based assessments.

Overall, the current results demonstrate that the fusion of generative and discriminative models within a single framework has the potential to significantly enhance breast thermography analysis. This approach contributes to develop scalable, AI-assisted tools for early screening in clinical and resource-limited settings.

6   Conclusion and Future Work

This study presented an end-to-end diagnostic framework that integrates Generative Adversarial Networks (GANs) and Convolutional Neural Networks (CNNs) for the early detection of breast anomalies using thermal imaging. Addressing the challenges of data scarcity, variability, and limited standardization in infrared thermography, the proposed system combines image preprocessing, synthetic data generation, and automated classification into a unified and reproducible pipeline. By implementing a CycleGAN model for unpaired image translation, the framework successfully generated realistic thermal images that enhanced the training dataset. The inclusion of these synthetic samples significantly improved the performance of deep learning classifiers, with EfficientNetB0 achieving an accuracy of 97.3% and an AUC of 0.977. These results surpass benchmarks reported in prior studies and validate the clinical relevance of the approach. Unlike prior works that focus solely on image synthesis or segmentation, the methodology explored in this paper highlights the potential of GAN-generated data when seamlessly integrated into classification pipelines. Furthermore, the modular architecture of the system allows for scalability and adaptation to other imaging modalities or diagnostic contexts. Despite its strengths, the study acknowledges limitations such as dataset size and the need for clinical validation. Future work will focus on expanding the dataset, incorporating domain-specific conditioning in GANs, and evaluating the system in real-world screening environments in collaboration with medical professionals. Overall, the findings support the feasibility of deploying deep learning systems based on synthetic thermography as complementary diagnostic tools, particularly in resource-constrained settings. The integration of generation and classification in a single diagnostic flow opens new avenues for AI-assisted breast cancer screening with increased accuracy, interpretability, and accessibility.

Acknowledgments. This work was funded by the Cátedra Fundación ASISA - Universidad Europea de Madrid under the 2024 Research Grants Program for Biomedical and Health Sciences projects, titled "Development and clinical use of an early diagnosis system and precise anomaly detection with deep learning algorithms and thermography."

Disclosure of Interests. The authors declare no conflicts of interest that could have influenced the results presented or their interpretation. Specifically, no author has received personal compensation, contracts, or financial benefits related to private entities that could be linked to the development, application, or commercialization of the algorithms, systems, or technologies addressed in this work.

Author Declaration on the Use of AI Tools. In compliance with Springer LNCS ethical guidelines, the authors declare that no part of this manuscript was generated by large language models (LLMs) or AI tools such as ChatGPT or GPT-4. All content, including scientific analysis, methodology, and conclusions represent original human work. Only minor language editing tools (Grammarly, DeepL Write) were used for stylistic refinement without altering scientific meaning. No AI was employed for research design, figure creation, or code generation.

References

1. Bray, F., et al.: Global cancer statistics 2022: GLOBOCAN estimates of incidence and mortality worldwide for 36 cancers in 185 countries. CA Cancer J. Clin. **74**(3), 229–263 (2024). https://doi.org/10.3322/caac.21834
2. World Health Organization: Breast cancer [Fact sheet] (2024). https://www.who.int/news-room/fact-sheets/detail/breast-cancer. Accessed July 26 2025
3. Azeredo-da-Silva, A.F., et al.: Health care accessibility and mobility in breast cancer: a Latin American perspective. BMC Health Serv. Res. **24**(1), 764 (2024). https://doi.org/10.1186/s12913-024-11222-6
4. Bekteshi, V., Schootman, M.: Bridging gaps in breast cancer screening: a comparative study of Mexican women in U.S. rural and urban areas. J. Prim. Care Community Health **15**, 21501319241295916 (2024). https://doi.org/10.1177/21501319241295916
5. Sala, D.C.P., Tanaka, O.Y., Luz, R.A., et al.: Barriers and facilitators of the implementation of mammography screening in the Brazilian public health system: scoping review. BMC Public Health **25**, 1659 (2025). https://doi.org/10.1186/s12889-025-22889-9
6. Ng, E.Y.K.: A review of thermography as promising non-invasive detection modality for breast tumor. Int. J. Therm. Sci. **48**(5), 849–859 (2009). https://doi.org/10.1016/j.ijthermalsci.2008.06.015
7. Singh, D., Singh, A.K.: Role of image thermography in early breast cancer detection: past, present and future. Comput. Methods Programs Biomed. **183**, 105074 (2020). https://doi.org/10.1016/j.cmpb.2019.105074
8. Liu, Q., et al.: Infrared thermography in clinical practice: a literature review. Eur. J. Med. Res. **30**(1), 33 (2025). https://doi.org/10.1186/s40001-025-02278-z
9. Goñi-Arana, A., et al.: Breast thermography: a systematic review and meta-analysis. Syst. Rev. **13**, 295 (2024). https://doi.org/10.1186/s13643-024-02708-9
10. Jalloul, R., et al.: Enhancing early breast cancer detection with infrared thermography: a comparative evaluation of deep learning and machine learning models. Technologies **13**(1), 7 (2025). https://doi.org/10.3390/technologies13010007

11. Ammer, K., Ring, E.F.J.: Standard procedures for infrared imaging in medicine. In: Ring, F. (ed.) Infrared Imaging: A Clinical Perspective, pp. 32.1–32.14. CRC Press (2012)

12. Korczak, I., et al.: Numerical prediction of breast skin temperature based on thermographic and ultrasonographic data in healthy and cancerous breasts. Biocybern. Biomed. Eng. **40**(4), 1680–1692 (2020). https://doi.org/10.1016/j.bbe.2020.10.007

13. Kakileti, S.T., Manjunath, G., Schwartz, R.G., Frangi, A.F. (Eds.). Artificial Intelligence over Infrared Images for Medical Applications: Second MICCAI Workshop, AIIIMA Proceedings. Springer Nature (2023). https://doi.org/10.1007/978-3-031-76584-1

14. Kakileti, S.T., Manjunath, G., Eddie, E.Y., Schwartz, R.G.: Artificial Intelligence Over Infrared Images for Medical Applications. AIIIMA Proceeding, Springer Nature (2024). https://doi.org/10.1007/978-3-031-76584-1

15. Goodfellow, I., et al.: Generative adversarial networks. In: Advances Neural Information Processing System **27** (2014). https://doi.org/10.48550/arXiv.1406.2661

16. Yi, X., Walia, E., Babyn, P.: Generative adversarial network in medical imaging: a review. Med. Image Anal. **58**, 101552 (2019). https://doi.org/10.48550/arXiv.1809.07294

17. LeCun, Y., Bengio, Y., Hinton, G.: Deep learning. Nature **521**(7553), 436–444 (2015). https://doi.org/10.1038/nature14539

18. Litjens, G., et al.: A survey on deep learning in medical image analysis. Med. Image Anal. **42**, 60–88 (2017). https://doi.org/10.1016/j.media.2017.07.005

19. Shorten, C., Khoshgoftaar, T.M.: A survey on image data augmentation for deep learning. J. Big Data **6**(1), 1–48 (2019). https://doi.org/10.1186/s40537-019-0197-0

20. Chlap, P., et al.: A review of medical image data augmentation techniques for deep learning applications. J. Med. Imaging Radiat. Oncol. **65**(5), 545–563 (2021). https://doi.org/10.1111/1754-9485.13261

21. Shojaedini, S.V., et al.: Generative adversarial network: a statistical-based deep learning paradigm to improve detecting breast cancer in thermograms. Med. Biol. Eng. Comput. **62**(5), 1077–1087 (2024). https://doi.org/10.1007/s11517-023-02989-7

22. Bossuyt, P.M.M., et al.: STARD 2015: an updated list of essential items for reporting diagnostic accuracy studies. BMJ **351**, h5527 (2015). https://doi.org/10.1136/bmj.h5527

23. Suganthi, S.S., Ramakrishnan, S.: Anisotropic diffusion filter based edge enhancement for segmentation of breast thermogram using level sets. Biomed. Signal Process. Control **10**, 128–136 (2014). https://doi.org/10.1016/j.bspc.2014.01.008

24. Venkatachalam, N., et al.: Enhanced segmentation of inflamed ROI to improve the accuracy of identifying benign and malignant cases in breast thermogram. J. Oncol. **2021**, 5566853 (2021). https://doi.org/10.1155/2021/5566853

25. Garia, L., Muthusamy, H.: Dual-tree complex wavelet pooling and attention-based modified U-Net architecture for automated breast thermogram segmentation and classification. J. Imaging Inform. Med. **38**, 887–901 (2024). https://doi.org/10.1007/s10278-024-01239-y

26. Aggarwal, A., Mittal, M., Battineni, G.: Generative adversarial network: an overview of theory and applications. Int. J. Inf. Manag. **60**, 100004 (2021). https://doi.org/10.1016/j.jjimei.2020.100004

27. Zhu, J.Y., Park, T., Isola, P., Efros, A.A.: Unpaired image-to-image translation using cycle-consistent adversarial networks. In: Proceedings of the IEEE Interna-

tional Conference on Computer Vision, pp. 2223–2232 (2017). https://doi.org/10.48550/arXiv.1703.10593

28. Heusel, M., et al.: GANs trained by a two time-scale update rule converge to a local Nash equilibrium. In: Advances in Neural Information Processing Systems, pp. 6626–6637 (2017)

29. Bińkowski, M., et al.: Demystifying MMD GANs. In: International Conference Learning Represent (2018)

30. Alzahrani, R.M., et al.: Early breast cancer detection via infrared thermography using a CNN enhanced with particle swarm optimization. Sci. Rep. **15**, 25290 (2025)

31. Tan, M., Le, Q.: EfficientNet: rethinking model scaling for convolutional neural networks. In: Proceedings of the 36th International Conference on Machine Learning, pp. 6105–6114. PMLR (2019)

32. He, K., et al.: Deep residual learning for image recognition. In: Proceedings of the IEEE Conference Computing Vision Pattern Recognition, pp. 770–778 (2016)

33. Al Husaini, M.A.S., et al.: Thermal-based early breast cancer detection using inception V3, inception V4 and modified inception MV4. Neural Comput. Appl. **34**(1), 333–348 (2022)

34. Mitchell, A.J.: Sensitivity × PPV is a recognised test called the Clinical Utility Index (CUI+). Eur. J. Epidemiol. **26**(3), 251–252 (2011)

35. Petrini, D.G.P., et al.: Breast cancer diagnosis in two-view mammography using end-to-end trained EfficientNet-based convolutional network. IEEE Access **10**, 77723–77746 (2022)

36. Shen, L., et al.: Deep learning to improve breast cancer detection on screening mammography. Sci. Rep. **9**(1), 12495 (2019)

37. Alshehri, A., AlSaeed, D.: Breast cancer detection in thermography using convolutional neural networks (CNNs) with deep attention mechanisms. Appl. Sci. **12**(24), 12922 (2022)

38. Govindaraju, B., Kakileti, S.T.: Generative artificial intelligence approaches for synthesizing high-fidelity breast thermal images. In: Artificial Intelligence over Infrared Images for Medical Applications. Proceedings of the Third International Conference, AIIIMA 2024, Virtual Event, 9 November 2024. LNCS, vol. 15279, pp. 33–43 (2024). https://doi.org/10.1007/978-3-031-76584-1_4

39. Silva, C.E.C., Conci, A.: About the validity of using DCGANs for data augmentation in breast thermography segmentation. In: Artificial Intelligence over Infrared Images for Medical Applications. Proceedings of the Third International Conference, AIIIMA 2024, Virtual Event, 9 November 2024. LNCS, vol. 15279, pp. 44–56 (2024). https://doi.org/10.1007/978-3-031-76584-1_5

Interpretable Breast Cancer Risk Stratification Using Statistical Feature Engineering on Thermal Images

Chandrima Hazra(✉) 📵, Mini Jayan 📵, and J. S. Saleema 📵

Department of Statistics and Data Science, Christ (Deemed to be University), Bangalore, Karnataka, India

Abstract. This research solves the "black box" problem of AI implementation in imaging by introducing a transparent, statistically grounded approach to breast cancer risk stratification via infrared thermography without compromising performance. Using the public DMR-IR dataset, statistical feature engineering was applied to training data by extracting first- and second-order statistical features. After ensuring non-normality with a Shapiro-Wilk test, feature significance was established with the Mann-Whitney U test. LASSO regularization selected the five most predictive features: mean, standard deviation, kurtosis, correlation, and energy. To counteract class imbalance, SMOTE was applied, and two machine learning models—logistic regression and random forest (classifier)—were trained on the balanced data and then evaluated on an unseen test dataset. Reporting an AUC of 0.98 over logistic regression's 0.96 reflects stringent statistical feature engineering and great generalization, creating a strong and interpretable model for breast cancer diagnosis in thermal images, and instills more clinical confidence in AI-based diagnostic systems.

Keywords: Breast Cancer · Infrared Thermography · Statistical Feature Engineering · Interpretable AI · Machine Learning · SMOTE · LASSO

1 Introduction

Breast cancer remains a significant global health challenge, standing as the most commonly diagnosed cancer and a leading cause of cancer-related mortality among women worldwide [1]. The cornerstone of improving patient outcomes lies in early and accurate detection. While mammography has been the standard for screening for decades and has been instrumental in reducing mortality [2], its limitations are well-documented. The procedure involves ionizing radiation, can cause significant discomfort, and, most critically, exhibits reduced diagnostic sensitivity in the nearly 40% of women with dense breast tissue, a known risk factor for the disease. These challenges underscore the pressing clinical need for safe, accessible, and effective adjunctive screening modalities.

Infrared (IR) thermography has emerged as a promising non-invasive, non-contact, and radiation-free imaging technique that addresses many of these limitations. Unlike

S. T. Kakileti et al. (Eds.): AIIIMA 2025, LNCS 16308, pp. 65–79, 2026.
https://doi.org/10.1007/978-3-032-10990-3_5

mammography, which visualizes anatomical structures, thermography captures physiological information by measuring the infrared radiation emitted from the skin's surface. The diagnostic premise is rooted in the distinct physiological characteristics of malignant tumors, which are often characterized by an increased metabolic rate and angiogenesis—the formation of new blood vessels [3]. This heightened cellular activity leads to localized elevations in temperature, providing a functional "heat map" of the breast that can reveal abnormalities.

The integration of Artificial Intelligence (AI) offers the potential to automate the analysis of these complex thermal patterns. In recent years, the field has been dominated by deep learning models, which have achieved high accuracy but at the cost of interpretability. This "black box" nature is a major impediment to clinical adoption, as clinicians are hesitant to trust a recommendation without understanding its basis, raising critical issues of trust, safety, and accountability.

This paper directly addresses this critical research gap. We present a transparent, statistically-grounded pipeline for breast cancer risk stratification that prioritizes interpretability without sacrificing performance. By focusing on rigorous statistical feature engineering—a targeted form of radiomics—we develop a "glass box" model that is both highly accurate and entirely understandable. Our contribution is a comprehensive methodology that validates the power of an interpretable-by-design approach, demonstrating that high performance and clinical trustworthiness can coexist.

2   Related Works

The application of computational analysis to breast thermography is a well-established field that has evolved significantly, moving from foundational feature engineering to complex deep learning systems. Our work is positioned at the intersection of classical statistical methods and modern machine learning, with a strong emphasis on interpretability.

2.1   Foundational Approaches in Feature Engineering

Early research focused on quantifying thermal patterns using handcrafted features. Many initial studies, such as the foundational work by Ng and Collins [5], centered on thermal asymmetry between the left and right breasts as a primary indicator of abnormality. This principle, which posits that malignancies disrupt normal thermal symmetry, became a cornerstone for many subsequent feature engineering approaches [6]. Building on this, researchers began incorporating a wider range of statistical descriptors. Foundational studies, such as the work by Acharya et al. [7], utilized first-order statistics and texture features with a Support Vector Machine (SVM) classifier, successfully demonstrating the viability of this feature-based approach for distinguishing normal from abnormal thermograms. Similarly, the work by Khodadadi and Nazem [12] and Abdel-Nasser et al. [8] successfully applied machine learning models to features derived from the DMR-IR dataset, achieving high classification accuracy and establishing strong benchmarks for feature-based methods. These foundational studies proved that statistical representations of thermal images contain significant diagnostic information.

2.2 Advancements in Texture and Statistical Analysis

As the field matured, the sophistication of feature extraction grew. The Gray-Level Co-occurrence Matrix (GLCM), which we utilize in our work, became a standard for capturing image texture. Studies such as the one by Abdel-Nasser et al. [8] specifically leveraged texture analysis methods, including GLCM, for thermogram classification, highlighting their power. Other researchers explored different feature families, such as Gray-Level Run Length Matrix (GLRLM) features [8] and variants of Local Binary Patterns (LBP), as demonstrated in the work by Madhavi and Bobby [9], to capture different aspects of thermal texture.

The success of these varied feature sets underscores a key theme: that the translation of thermal patterns into a structured feature space is a powerful and flexible methodology. Our work builds directly on this paradigm by combining multiple feature types and employing a rigorous selection process.

2.3 The Rise of Deep Learning in Thermography

With the advent of deep learning, many researchers shifted focus to end-to-end models, particularly Convolutional Neural Networks (CNNs). Architectures like VGG-16, ResNet, and Inception have been widely applied to thermal image classification [10–12]. These models have the advantage of learning relevant features automatically from raw pixel data, often achieving state-of-the-art performance [13, 14]. For instance, recent deep learning approaches, such as the attention-based CNNs developed by Alshehri and AlSaeed [15] and the compact multi-architecture strategy proposed by Attallah [16] on the DMR-IR dataset, have reported accuracies exceeding 90%. However, this performance comes at the cost of interpretability, creating the "black box" problem that our research directly aims to address.

2.4 The Quest for Explainable AI (XAI)

To mitigate the "black box" issue, a subfield of Explainable AI (XAI) has emerged. Techniques like Gradient-weighted Class Activation Mapping (Grad-CAM) are often used to generate heatmaps that visualize which parts of an image a CNN is focusing on [17, 18]. While valuable, these post-hoc explanation methods provide an approximation of the model's behavior rather than a direct insight into its logic. Our work is fundamentally different, as we build a model that is *inherently interpretable* by design, a concept often referred to as "glass box" AI.

2.5 Addressing Data Challenges in Medical Imaging

A common challenge in medical imaging is dataset imbalance, where healthy samples far outnumber pathological ones, or vice versa. This issue has been widely addressed in the literature, with the Synthetic Minority Over-sampling Technique (SMOTE), which we use, being one of the most popular and effective methods [19–21]. The successful application of SMOTE in our pipeline aligns with best practices established across the broader field of medical image analysis.

3   Methodology

Our process was straightforward: we started with the raw thermal images, extracted and selected the most important statistical clues, and used them to build our models. The crucial last step was a final exam, testing the models on data they had never encountered to see how well they performed.

3.1   Dataset

This study utilized the public Database for Mastology Research with Infrared Image (DMR-IR) [4]. The dataset is pre-divided into a **training set** and a separate **test set**. Our entire feature engineering, selection, and model training process was conducted exclusively on the training data. The test set was held out and used only for the final, unbiased evaluation of the trained models (Fig. 1).

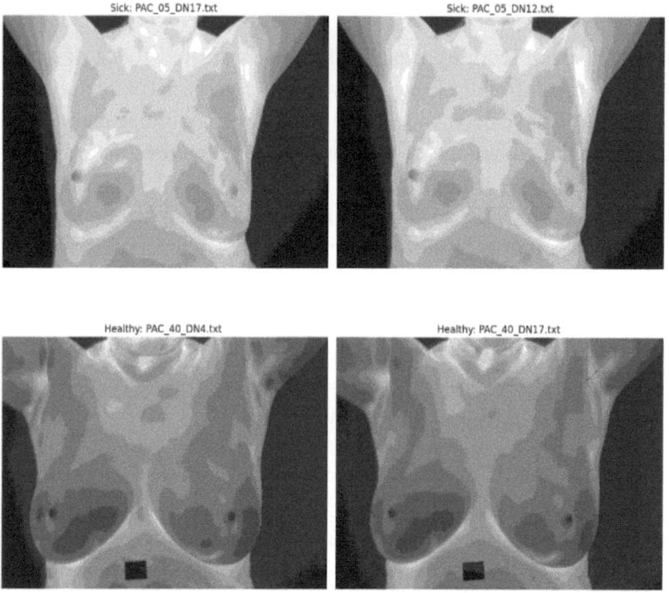

Fig. 1. Examples of thermal images from the DMR-IR dataset.

These images show us the patterns of a sick (top two images) and a healthy (bottom two) patient. Distinct hotspots with irregular and asymmetric heat patterns are shown on the sick images due to angiogenesis and the higher metabolic rate of the tumor. Still, healthy tissue typically shows cooler, uniform, and symmetrical heat patterns.

3.2   Feature Engineering

Before thinking about analysis, we had to process our raw materials—the images. A computer doesn't see a picture; it needs a blueprint with measurements. So, we went through each image and took eight specific measurements, like average temperature, texture, etc. This gave us a structured dataset from which to build. These eight "measurements" can be grouped like this:

First-Order (Intensity) Statistics. These features extracted from the images describe the global distribution of pixel intensities:

- **Mean:** It is the average temperature value of the whole image.
- **Standard Deviation:** A measure of thermal heterogeneity.
- **Skewness:** Quantifies the asymmetry of the temperature distribution.
- **Kurtosis:** Measures the "tailedness" of the distribution of the features.

Second-Order (Texture) Statistics. These features are a bit more advanced and look at how pixels relate to their neighbors. They help us understand the texture of the heat map. We get these by analyzing a Gray-Level Co-occurrence Matrix (GLCM), a formal way of tallying how often different temperature values appear next to each other.

- **Contrast:** Measures local intensity variations.
- **Correlation:** Measures the linear dependency of grey levels.
- **Energy:** Quantifies textural uniformity.
- **Homogeneity:** Measures the closeness of the distribution of elements in the GLCM to the diagonal.

3.3   Data Preprocessing and Feature Selection

Exploratory Data Analysis (EDA). Initial EDA on the engineered feature set revealed differences in the distributions between the 'healthy' and 'sick' classes (Fig. 2).

In the box plot, we can see that some features are stronger individual predictors than others, like the mean feature, which shows extreme separation, whereas the standard deviation (blood pressure fluctuation) makes significant overlap. Still, we consider all features because our model needs a complete clinical picture to make the most accurate diagnosis.

Statistical Significance Testing. A Shapiro-Wilk test was performed on all features, which rejected the null hypothesis of normality ($p < 0.05$). Just because the features for the "healthy" and "sick" groups looked different in our initial charts didn't mean it was a statistical certainty. We had to be sure these differences were fundamental and not just a fluke of the data.

To do this, we used a specific statistical test called the Mann-Whitney U test. This test is perfect for comparing two groups and telling you if the differences you see are significant. The results from this test confirmed that our chosen features behaved differently for the healthy and sick patient images.

Correlation Analysis. A correlation matrix was generated to assess multicollinearity among the features. Highly correlated features can introduce redundancy and instability in some models (Fig. 3).

The color of each square tells us if two features have a strong or weak connection.

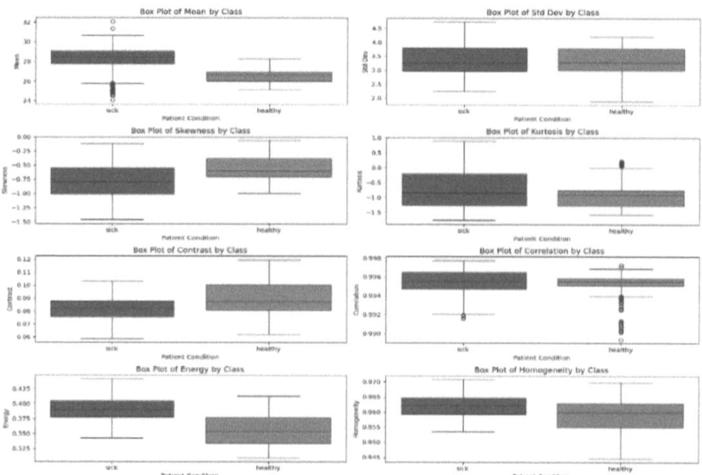

Fig. 2. Box plots of the features showing the range of values, helping us quickly identify which characteristics are most different between the 'healthy' and 'sick' classes.

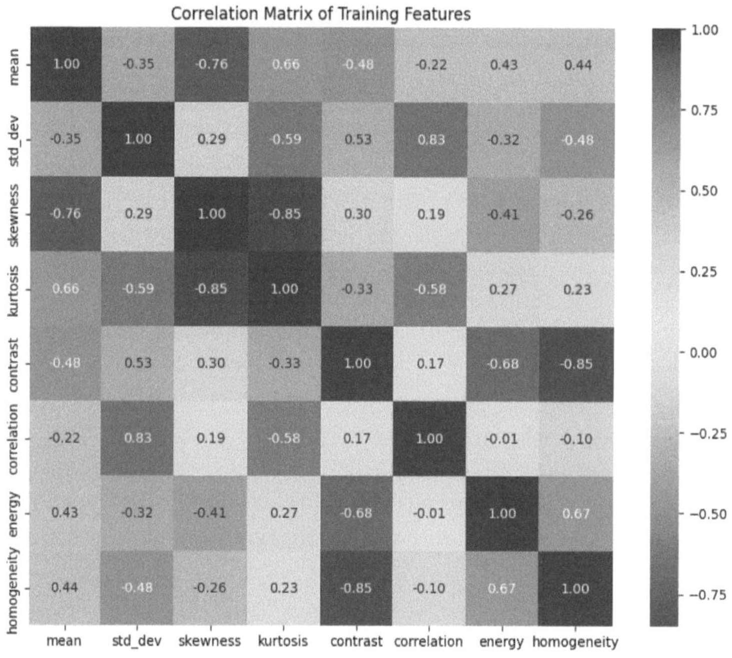

Fig. 3. A heatmap that visually shows how strongly our eight statistical features are related.

LASSO Regularization for Feature Selection. Looking at features in isolation is a good start, but it doesn't tell the whole story. The real challenge is that features often overlap, and a basic test can't pick the winning combination for a predictive model. We chose LASSO logistic regression specifically to solve this. Think of it like a filter that automatically keeps the most important variables and discards the noise. This step was crucial for building an accurate and reliable model.

LASSO works by adding a penalty to the model based on the absolute value of the feature coefficients. As this penalty (controlled by a hyperparameter, alpha) increases, the model is forced to shrink the coefficients of less essential features towards zero. Features whose coefficients are reduced to exactly zero are effectively removed from the model.

This process is visualized in Fig. 4. The x-axis represents the regularization strength (alpha), and the y-axis shows the standardized coefficient value for each of our eight initial features. As we move from left to right (increasing the penalty), we can observe the coefficient paths for each feature. The paths for correlation, homogeneity, and skewness are among the first to be forced to zero, indicating they are the least essential predictors in the presence of the others. In contrast, features like mean and std_dev retain significant, non-zero coefficients for much longer, demonstrating their strong predictive power.

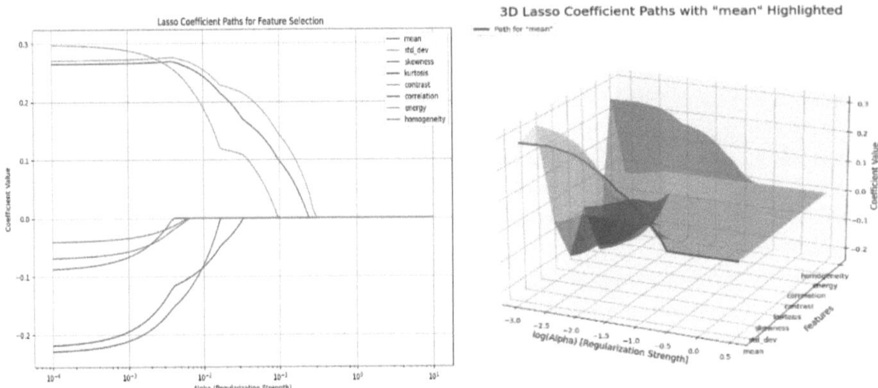

Fig. 4. Lasso Coefficient Paths for Feature Selection.

As the regularization strength (α) increases, the coefficients of less predictive features are shrunk to zero, effectively removing them from the model.

Based on this analysis, we selected the optimal set of features whose coefficients remained non-zero at a tuned regularization strength. This resulted in the final selection of **five features** for our models: **mean, std_dev, kurtosis, contrast, and energy.** This automated and data-driven approach ensures that our final models are both robust and parsimonious.

Handling Class Imbalance with SMOTE. The training data presented a clear imbalance, with a disproportionate number of samples in the 'sick' category. This is risky because a model could achieve high apparent accuracy by simply defaulting to the 'sick' prediction. This would make it unreliable for detecting the 'healthy' class, which is often the group of greater interest.

To fix this data imbalance, we used a technique called SMOTE. Instead of just duplicating the 'healthy' samples, SMOTE creates new, realistic ones. It works by picking a sample from the 'healthy' group, looking at its closest neighbors (other 'healthy' samples), and then generating a new, synthetic sample somewhere between them. This process effectively beefed up the minority class, giving us a much more balanced dataset to train the model.

This process was applied only to the training portion of our data to prevent data leakage into the test set. The effect of this technique is visualized in Fig. 5, where the feature space is plotted using its first two principal components. The plot on the left shows the original imbalanced data, while the plot on the right shows the balanced dataset after SMOTE has generated new synthetic samples for the 'healthy' class.

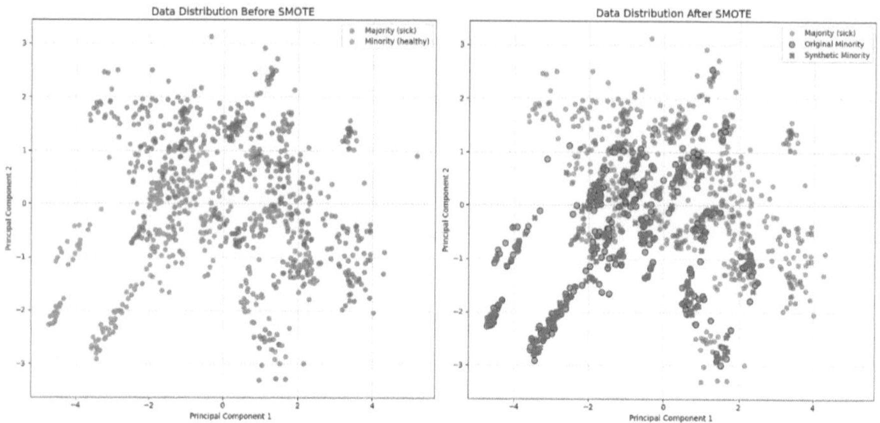

Fig. 5. Distribution of data points before (left) and after (right) applying SMOTE.

In Fig. 5, the green points represent the synthetic samples generated for the minority ('healthy') class, creating a balanced dataset for model training.

The positive impact of this balancing technique on model performance was confirmed by analyzing the learning curves of our classifiers. As shown in Fig. 6, the validation accuracy for both the logistic regression and the random forest models was much higher after training on the SMOTE-balanced data, demonstrating improved generalization.

The validation accuracy for both models (dashed lines) improved after training on the balanced dataset.

3.4 Modeling and Evaluation

With our data prepped and balanced, it was time for us to build and train our models. It's crucial that a model be tested on data it has never seen before—otherwise, you don't know if it has actually learned or just memorized the answers.

To do this, we split our initial training data into two parts:

- **80%** was used as the **training set**, which the model would study to learn the patterns.
- The remaining **20%** was set aside as the **validation set**. This data was kept separate and used as a final exam to get an honest grade on how well the model performed on new information.

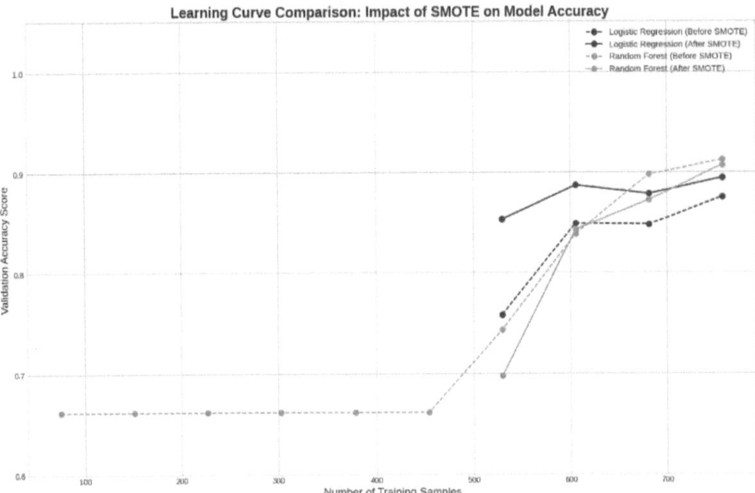

Fig. 6. Learning curve comparison showing the impact of SMOTE.

The SMOTE-balanced 80% set was used to train two distinct machine learning classifier models:

1. **Logistic Regression:** A linear, highly interpretable baseline model.
2. **Random Forest:** A robust, nonlinear ensemble model.

The models were first evaluated on the 20% validation set to make a quick check and shakedown test. But finally, the models retrained on the entire original training set were assessed on the completely unseen test data, along with a 5-fold cross-validation splitter to determine the trade-off between their real-world generalization performance and a stable estimate of the models' learning. Evaluation was based on accuracy, confusion matrices, precision, recall, and F1-score.

4 Results

Our feature selection process worked perfectly, narrowing the eight initial clues down to the five that did most of the heavy lifting: **mean, standard deviation, kurtosis, correlation**, and **energy**.

With our top five features selected, we trained our models and then moved on to the most crucial step: the final test of our model. We unleashed the trained models on the validation data—the portion they had never seen before—to see how they would perform in a real-world scenario.

4.1 Model Performance on Unseen Data

We have reached the point where both models did a good job of telling the difference between healthy and sick cases. However, the Random Forest model stood out as having better performance. This shows that the advantage of using such a model is that it can learn complex and less obvious patterns in the data.

We can see a full breakdown of the performance numbers for both models in Table 1.

Table 1. Performance Metrics of Final Models on the Unseen Test Set.

Model	Accuracy	Precision	Recall	F1-score	AUC
Logistic Regression	0.71	0.70 (sick), 1.00 (healthy)	1.00 (sick), 0.13 (healthy)	0.82 (sick), 0.24 (healthy)	0.96
Random Forest	0.96	0.94 (sick), 0.98 (healthy)	1.00 (sick), 0.88 (healthy)	0.97 (sick), 0.93 (healthy)	0.98

The confusion matrices below provide a granular view of the classification results for each model on the test set (Fig. 7).

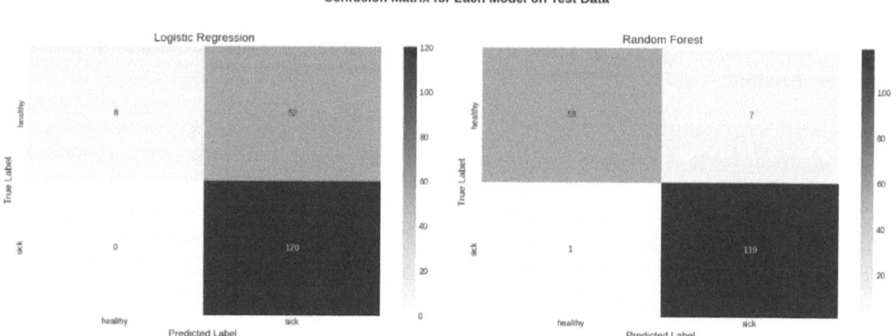

Fig. 7. Confusion matrices for the logistic regression (left) and random forest (right) models on the unseen test set.

4.2 Final Evaluation on Unseen Test Data

For the final round of evaluation, we brought out a completely new dataset that neither model had ever encountered before. This was the ultimate test to see how well they could apply their knowledge to new situations.

Once again, the **Random Forest** model came out on top. It had the highest accuracy (96%), which is very important: it showed superior "generalization." This means it's not just good at recognizing patterns it has already studied, but it is also more reliable and accurate when making a judgment on a completely fresh case (Fig. 8).

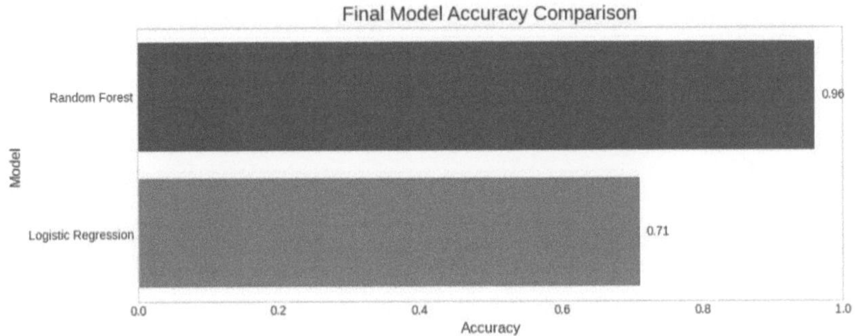

Fig. 8. Comparison of model accuracies on the unseen test dataset.

The final features extracted from the random forest model were determined by its feature importance score and are presented below (Fig. 9).

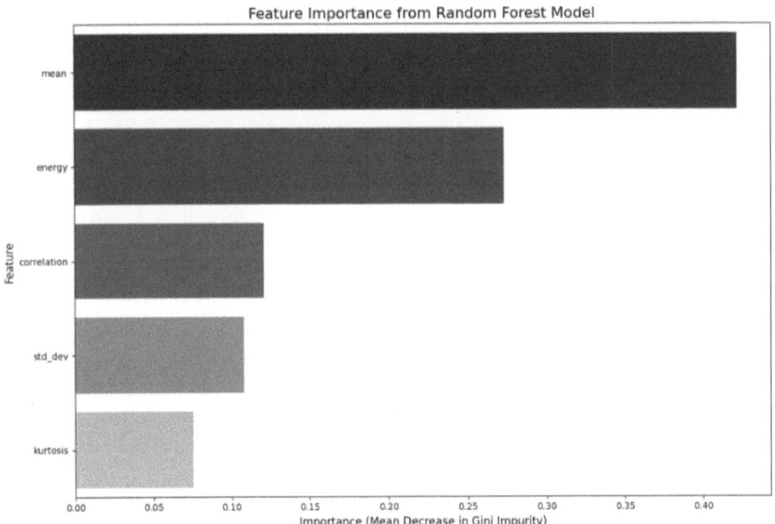

Fig. 9. Features extracted from the Random Forest model.

The random forest model with an ensemble of trees, which is inherently nonlinear, performs excellently because in our data, it helps to create highly complex and flexible decision boundaries that capture the proper relationship between features. The model's real power comes from its ability to see how features work as a team. Like through a single decision tree, it might learn that if the mean temperature is high and contrast is low, the patient is likely sick. Hence, it can understand complex patterns because simpler models and univariate tests alone can't do this in medical problems. Our models are not going through any data leakage because they consistently perform, first in the five-fold cross-validation in the training set and then in a completely different held-out test set.

Also, we prevent data leakage by correctly fitting all preprocessors like StandardScaler, Lasso Selector, and SMOTE balancing techniques on the training data, ensuring it only sees the training data within each cross-validation fold.

5 Discussion

Our study successfully designed and validated a transparent, statistically-driven pipeline for breast cancer classification using thermal images, achieving an outstanding AUC of 0.98 with a Random Forest model on unseen test data. This section discusses the interpretation of these findings, compares our methodology to recent techniques, and outlines the clinical implications of our work.

5.1 Interpretation of Key Findings

The core success of our pipeline lies in its rigorous feature selection process. LASSO regularization identified a parsimonious set of five features—mean, standard deviation, kurtosis, correlation, and energy—as the most salient predictors. This finding is clinically significant. The importance of mean and std_dev aligns with the foundational theory of thermography that malignant regions are both hotter and exhibit greater thermal heterogeneity. The inclusion of kurtosis suggests that the presence of extreme thermal outliers is a key diagnostic indicator, while the texture features (correlation, energy) confirm that the spatial arrangement of thermal patterns, not just their intensity, is crucial for accurate classification. The high performance of both the linear Logistic Regression and the non-linear Random Forest models underscores the rich, predictive information contained within this small, interpretable feature set.

5.2 Comparison with Recent Techniques

Our methodology, which relies on extracting statistical features from images, can be broadly categorized as a radiomics approach. It aligns with similar radiomics-based analyses, done by Gupta and Sharma [22], which also leveraged quantitative features from thermal images for prediction. However, our approach differs in its specific emphasis on creating a minimal, highly interpretable "glass box" model. While many radiomics studies employ a large number of features, our use of LASSO actively prioritizes a small, parsimonious set, making the final model's logic easier for a clinician to understand and trust.

Our approach stands in contrast to the end-to-end deep learning methods that have become prevalent in the field. For instance, advanced techniques such as the 3D Convolutional Neural Network proposed by Baffa et al. [23] have been developed to analyze dynamic breast infrared imaging, capturing spatiotemporal temperature changes. Such models have the advantage of learning complex features automatically from raw imaging data and can achieve very high performance. However, they operate as "black boxes," making it difficult to understand their decision-making process. Our Random Forest model, with its 98% AUC, achieves a performance that is highly competitive with these deep learning systems, but with the crucial added benefit of full transparency. We can precisely identify and rank the importance of the features driving the prediction (as shown in our Feature Importance results), which is a capability opaque models lack. Our work, therefore, demonstrates that the often-assumed trade-off between top-tier accuracy and interpretability may not be as stark as previously thought.

5.3   Strengths, Limitations, and Clinical Implications

The primary strength of this study is its transparency. By building an interpretable-by-design model, we offer a solution that can serve as a true clinical decision support tool, providing not just a prediction but the reasoning behind it. This is essential for fostering the clinical trust necessary for real-world adoption.

However, we acknowledge limitations that open avenues for future work. Our model is currently binary ('healthy' vs. 'sick'). As discussed in the Future Work section, a crucial next step is to develop a multi-class model to differentiate benign from malignant conditions. Furthermore, while our model performs exceptionally well on the DMR-IR dataset, external validation on data from different institutions is necessary to confirm its generalizability.

6   Conclusion

Ultimately, this research has shown the entire pipeline to create a clear, understandable system for analyzing breast cancer risk from thermal images.

By focusing on pulling out a few meaningful statistical clues from the images and then carefully comparing our models, we built a Random Forest model that is both easy to interpret and highly accurate. Achieving an outstanding AUC score of **0.96 or 0.98** on completely new data proves its reliability.

Ultimately, this work offers a powerful alternative to the "black box" approach. It lights a path forward for developing innovative diagnostic tools that doctors can trust, blending modern AI's power with the clarity that medicine demands.

Acknowledgments. We built our analysis using the DMR-IR dataset, which was generously made public by the team at Brazil's Federal Fluminense University. We sincerely thank them for their commitment to open science and for providing this valuable resource.

Disclosure of Interests. The authors have no competing interests to declare that are relevant to the content of this article.

References

1. World Health Organization: Breast cancer. https://www.who.int/news-room/fact-sheets/detail/breast-cancer. Accessed 06 Aug 2025
2. Marmot, M.G., et al.: The benefits and harms of breast cancer screening: an independent review. The Lancet **380**(9855), 1778–1786 (2012)
3. Gautherie, M.: Thermopathology of breast cancer: measurement and analysis of in vivo temperature and blood flow. Ann. N. Y. Acad. Sci. **335**(1), 383–415 (1980)

4. Borges, L., et al.: A new database for breast research with infrared images. In: Proceedings of the 28th SIBGRAPI Conference on Graphics, Patterns, and Images, pp. 314–321. IEEE (2015)

5. Ng, E.Y., Fok, S.C.: A framework for early discovery of breast tumor using thermography with artificial neural network. Breast J. 9(4), 341–343 (2003). https://doi.org/10.1046/j.1524-4741.2003.09425.x

6. Tsietso, D., Yahya, A., Samikannu, R.: A review on thermal imaging-based breast cancer detection using deep learning. Mob. Inf. Syst. 2022, 1–19 (2022). https://doi.org/10.1155/2022/8952849

7. Acharya, U.R., Ng, E.Y.K., Tan, J.H., et al.: Thermography based breast cancer detection using texture features and support vector machine. J. Med. Syst. 36, 1503–1510 (2012). https://doi.org/10.1007/s10916-010-9611-z

8. Abdel-Nasser, M., Moreno, A., Puig, D.: Breast cancer detection in thermal infrared images using representation learning and texture analysis methods. Electronics 8(1), 100 (2019). https://doi.org/10.3390/electronics8010100

9. Madhavi, V., Bobby, T.C.: Thermal imaging based breast cancer analysis using BEMD and uniform RLBP. In: 2017 Third International Conference on Biosignals, Images and Instrumentation (ICBSII), pp. 1–6. Chennai, India (2017). https://doi.org/10.1109/ICBSII.2017.8082268

10. Aggarwal, A., Alpana, Pandey, M.: Deep Learning based Breast Cancer Classification on Thermogram. 769–774 (2022). https://doi.org/10.1109/ICCCIS56430.2022.10037624

11. Mahmoud, H., Alharbi, A., Khafaga, D.: Breast cancer classification using deep convolution neural network with transfer learning. Intell. Autom. Soft Comput. 29, 803–814 (2021). https://doi.org/10.32604/iasc.2021.018607

12. Khodadadi, H., Nazem, S.: Improving cancer detection through computer-aided diagnosis: a comprehensive analysis of nonlinear and texture features in breast thermograms. PLoS ONE 20(5), e0322934 (2025). https://doi.org/10.1371/journal.pone.0322934

13. Alshehri, A., AlSaeed, D.: Breast cancer detection in thermography using convolutional neural networks (CNNs) with deep attention mechanisms. Appl. Sci. 12, 12922 (2022). https://doi.org/10.3390/app122412922

14. Zuluaga, J.P., Al Masry, Z., Benaggoune, K., Meraghni, S., Zerhouni, N.: A CNN-based methodology for breast cancer diagnosis using thermal images. Comput. Methods Biomech. Biomed. Eng. Imaging Biomechanics Vis. 9. 1–15 (2020). https://doi.org/10.1080/21681163.2020.1824685

15. Alshehri, A., AlSaeed, D.: Breast cancer detection in thermography using convolutional neural networks (CNNs) with deep attention mechanisms. Appl. Sci. 12(24), 12922 (2022). https://doi.org/10.3390/app122412922

16. Attallah, O.: A deep learning-driven CAD for breast cancer detection via thermograms: a compact multi-architecture feature strategy. Appl. Sci. 15(13), 7181 (2025). https://doi.org/10.3390/app15137181

17. Selvaraju, R.R., Cogswell, M., Das, A., Vedantam, R., Parikh, D., Batra, D.: Grad-CAM: visual explanations from deep networks via gradient-based localization. In: 2017 IEEE International Conference on Computer Vision (ICCV), pp. 618–626. Venice, Italy (2017). https://doi.org/10.1109/ICCV.2017.74

18. Samek, W., Montavon, G., Vedaldi, A., Hansen, L., Müller, K.-R.: Explainable AI: interpreting, explaining and visualizing deep. Learning (2019). https://doi.org/10.1007/978-3-030-28954-6

19. Chawla, N., Bowyer, K., Hall, L., Kegelmeyer, W.: SMOTE: synthetic minority over-sampling technique. J. Artif. Intell. Res. (JAIR) 16, 321–357 (2002). https://doi.org/10.1613/jair.953

20. Johnson, J.M., Khoshgoftaar, T.M.: Survey on deep learning with class imbalance. J Big Data 6, 27 (2019). https://doi.org/10.1186/s40537-019-0192-5

21. Ronneberger, O., Fischer, P., Brox, T.: U-Net: convolutional networks for biomedical image segmentation. In: Navab, N., Hornegger, J., Wells, W., Frangi, A. (eds.) Medical Image Computing and Computer-Assisted Intervention – MICCAI 2015. MICCAI 2015. Lecture Notes in Computer Science(), vol. 9351. Springer, Cham (2015). https://doi.org/10.1007/978-3-319-24574-4_28
22. Shrivastava, R., Kakileti, S.T., Manjunath, G.: Thermal radiomics for improving the interpretability of breast cancer detection from thermal images. In: Kakileti, S.T., et al. Artificial Intelligence over Infrared Images for Medical Applications and Medical Image Assisted Biomarker Discovery. MIABID AIIIMA 2022 2022. Lecture Notes in Computer Science, vol. 13602. Springer, Cham (2022). https://doi.org/10.1007/978-3-031-19660-7_1
23. de Freitas Oliveira Baffa, M., Lattari, L.G., Conci, A.: 3D convolutional neural networks for dynamic breast infrared imaging classification. In: Kakileti, S.T., Manjunath, G., Schwartz, R.G., Frangi, A.F. (eds.) Artificial Intelligence over Infrared Images for Medical Applications. AIIIMA 2023. Lecture Notes in Computer Science, vol. 14298. Springer, Cham (2023). https://doi.org/10.1007/978-3-031-44511-8_4

Functional Imaging-Guided Early Detection of Radiation Pneumonitis: A Multimodal Framework Integrating CT Radiomics and Infrared Thermography

Sotiris Raptis[1]([⊠]) [iD], Christos Ilioudis[2] [iD], and Kiki Theodorou[1] [iD]

[1] Medical Physics Department, Medical School, University of Thessaly, Larisa, Greece
`sraptis@uth.gr`
[2] Department of Information and Electronic Engineering, International Hellenic University (IHU), Thessaloniki, Greece

Abstract. Radiation pneumonitis (RP) is an early inflammatory reaction of lung parenchyma that arises following thoracic radiotherapy. Characteristically, RP can begin to manifest physiologically days to weeks before overt apparent radiological evidence manifests on CT scans. That gap of time gives rise to a diagnostic blind spot that blunts effective intervention. In this paper, we propose an entirely new multimodal AI framework that seeks to functionally identify RP in its earliest expression of manifestation-prior to anatomical observability-via an integration of CT-based radiomics together with surrogate infrared thermal indicators that are related to respiratory dynamics and thoracic surface temperature. Since there is no dataset of paired CT and IR imaging of RP patients to work with for this study, we construct a feature-level fusion dataset with CT instances being matched with thermal signs sourced from public IR databases of pneumonia, COVID-19, and respiratory function disorder. These thermal indicators simulated clinical symptoms of incipient RP stage in terms of thoracic asymmetry, impaired breathing cycles, and focal hyperthermia. This study should be interpreted as a feasibility analysis: we integrate CT radiomics with physiologically motivated surrogate infrared (IR) descriptors to support very-early functional detection of radiation pneumonitis (RP) within the subclinical, pre-radiological window. Robustness was assessed with stratified cross-validation, ablations (CT-only vs CT + IR), and two control checks—label-preserving IR permutation and leave-one-IR-dataset-out evaluation—together with probability calibration. The fused model outperformed the CT-only baseline (AUC 0.91 vs 0.79; sensitivity 0.88 vs 0.68; F1 0.86 vs 0.72). We avoid generalization claims and motivate prospective paired CT–IR acquisition.

Keywords: Radiation-induced pneumonitis · CT radiomics · Infrared imaging · Multimodal learning · Respiratory biomarkers

Supplementary Information The online version contains supplementary material available at https://doi.org/10.1007/978-3-032-10990-3_6.

1 Introduction

Radiation pneumonitis (RP) is an inflammatory subacute side effect that arises in a considerable percentage of patients undergoing thoracic radiotherapy, particularly in patients with lung or mediastinal malignancy [1]. RP generally appears 1–6 months after therapy and presents with nonspecific yet grave symptoms of dry cough, mild dyspnea, chest discomfort, low-grade pyrexia, and in extreme cases of respiratory distress [2]. RP represents an acknowledged but relatively frequent toxicity, for which early diagnosis remains elusive due to clinical symptoms generally appearing before conventional computed tomography (CT) scanning reveals corresponding radiological evidence. Radiomics has emerged as a non-invasive AI-facilitated means of deriving high-dimensional features from standard-of-care CT scans that now offer new promise for prediction and diagnostic potential in radiation medicine and oncology [3]. Some researchers have successfully applied CT-based radiomics to predict RP risk and correlate imaging biomarkers with dose distribution, lung tissue characteristics, and patient prognosis. Those models significantly engage in pre-treatment prediction or post-hoc confirmation of RP once structural modification has become discernible on a radiological basis. That time interval between an initial physiology RP onset and discernibility on CT remains under-analyzed and diagnostically relevant. It is under this "diagnostic blind spot" that functional imaging has the potential for new breakthroughs. Inflammatory processes, as with those that underlie RP, usually produce physiological alterations-e.g., regional hyperthermia, respiratory symmetry defects, or compromised thermal breathing dynamics-before anatomical damage ensues [4]. Infrared thermography (IR), being a contactless and passive technique that captures surface temperature distributions, has already demonstrated potential in detecting such alterations in various pulmonary diseases, including viral pneumonia, COVID-19, and respiratory function impairment [5, 6]. Thermal imaging has been revealed to detect thoracic temperature asymmetry, eventual breathing disorders at their nascent stages, and systemic inflammatory reactions that are not discernible with structural imaging [7]. Infrared thermal imaging has been extensively validated in other thoracic applications, most notably in breast cancer detection, where asymmetrical surface temperature and vascular changes have served as early non-invasive biomarkers. This precedent supports the hypothesis that surface-level thermographic signals-when functionally aligned-may also reflect early inflammatory processes such as those seen in radiation-induced pneumonitis, even in the absence of direct intraparenchymal imaging [8–10]. While thermal cues possess significant physiological relevance, to date there has not been an openly published dataset providing matched IR and CT imaging of RP patients. To address this shortcoming, this paper proposes a new multimodal learning approach that pairs patients' radiomics features with known RP outcomes from their CT images, combined with surrogate thermal features, derived from publicly available independent infrared pathology datasets, showing similar presentation of early RP. The surrogate features of surface thermal activity and respiratory motion correlate with the early RP physiological signs of dysventilation and focal inflammation. Our assumption is that this integration of structural (CT radiomics) and functional (IR surrogate) information at a feature level can enhance sensitivity in detecting early RP [11]. We do not intend to predict RP prior to its occurrence, but to recognize it earlier than current methods of imaging permit, in the pre-radiological phase of physiological inflammation. Using explainable methods of machine learning, we intend to understand the role of thermal features in RP classification performance and clarify their potential relevance to clinical care. Through integration of data-driven functional and anatomic descriptors, this work introduces an alternative perspective of radiotherapy toxicity tracking that has the potential to help guide prospective clinical applications of real-time infrared observation in oncologic applications.

2   Materials and Methods

This study presents a multimodal artificial intelligence (AI) pipeline that combines anatomical and functional image-derived features to find early-stage RP [12, 13]. The proposed approach integrates two distinct data types at the feature level: high-dimensional radiomic features taken from thoracic CT scans (structural descriptors) and surrogate thermal signatures taken from publicly available infrared (IR) datasets related to respiratory conditions (functional descriptors). Although CT-based radiomics can characterize lung tissue patterns associated with RP, these features often manifest after the inflammatory response has caused anatomical changes detectable by imaging. On the other hand, thermal imaging captures physiological cues-such as localized surface temperature elevation, breathing asymmetry, and fluctuating respiration-that may reflect the functional onset of RP inflammation, even when structural changes are not yet radiologically evident. Currently, no dataset contains both CT and IR data for the same RP patients, so direct multimodal training is not possible. To fix this, we suggest a proxy fusion strategy that simulates multimodal inputs by giving CT cases with matching RP outcome labels representative thermal signatures from respiratory IR datasets.

2.1   Overview – Conceptual Design and Fusion Framework

The proposed pipeline comprises the following sequential phases (see Fig. 1):

- CT Dataset Collection and Labeling

We use a curated CT dataset of lung cancer patients who had thoracic radiotherapy and whose RP + or RP- outcomes are known. These cases are the building blocks of the analysis.

- Radiomics Feature Extraction from CT Images

Using the PyRadiomics toolkit, we get a wide range of custom features that measure patterns of intensity, texture, and shape in the lung Region of Interest (ROI) [14]. These features show how different tissues are and how radiation affects them.

- Acquisition of Public Thermal Imaging Datasets

Independent IR datasets, such as those for COVID-19, pneumonia, and healthy breathing, are used to simulate the early functional signs of RP. Each sample gives off thermal patterns over time and space, like breathing asymmetry and temperature gradients.

- Thermal Feature Extraction

From each IR image or video, we extract physiological indicators such as the chest thermal asymmetry index, the average respiratory rate from thermal variation, and the amplitude variability across breathing cycles.

- Feature-Level Fusion

Each CT case is fused with a thermal feature vector that is functionally related: RP + CT scans are given IR features that are like RP (such as high asymmetry or unstable patterns), whereas RP-scans are matched with thermal features from cases that are physiologically normal. This creates a multimodal training dataset that is synthetic but physiologically realistic.

- Model Training and Interpretation

The combined CT + IR feature space is used to train a Random Forest classifier. SHAP (SHapley Additive Explanations) is used to achieve post hoc explainability, which enables us to measure the diagnostic contribution of thermal features in comparison to radiomics [15].

- Performance Evaluation

Cross-validation is used to assess Area Under the Curve (AUC), specificity, and sensitivity. To evaluate the functional modality's additional diagnostic value, the fused model is contrasted with a baseline that uses only CT [16].

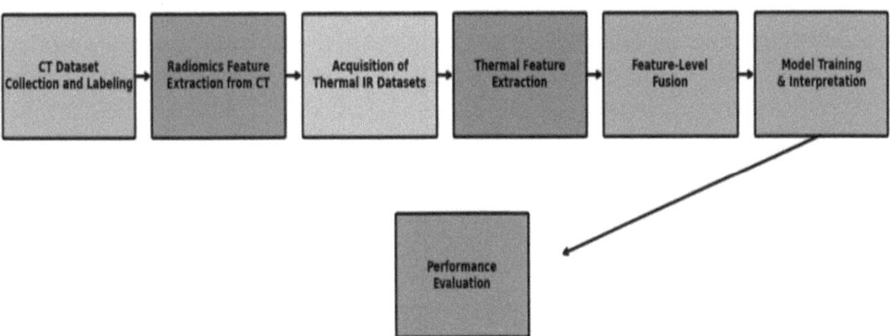

Fig. 1. The multimodal framework combining CT radiomics with surrogate infrared thermal features.

2.2   CT Dataset and Radiomics Feature Extraction

The structural foundation of our multimodal framework is based on thoracic computed tomography imaging of patients who underwent radiotherapy (RT) for lung cancer. The objective of this step is to extract a robust set of high-dimensional, handcrafted radiomics features that capture the morphological and textural characteristics of lung tissue, enabling downstream learning of subtle anatomical correlates of RP. We used a carefully selected external validation dataset of 120 anonymized CT scans that were gathered from patients with non-small cell lung cancer (NSCLC) who underwent definitive thoracic radiation therapy in the past [17–19]. Among the requirements for inclusion were:

- Completion of full RT regimen
- Availability of diagnostic-quality CT scans ($\leq$ 5 mm slice thickness)
- Clinical documentation of RP status within a 6-month post-RT window
- Cases were manually labeled by expert radiation oncologists as either:
- RP positive (RP +): Based on follow-up CT findings showing ground-glass opacities, consolidation patterns, and/or clinical confirmation (n = 60)
- RP negative (RP-): No imaging or clinical evidence of RP during the follow-up period (n = 60)

This 1:1 balanced design mitigates bias and enhances statistical stability during model training and evaluation. According to the guidelines provided by the Image Biomarker Standardization Initiative (IBSI), all CT scans underwent standardized preprocessing before feature extraction:

- Resampling to isotropic voxel spacing (1.0 $\times$ 1.0 $\times$ 1.0 mm3) using B-spline interpolation
- Intensity normalization to the [-1000, 400] Hounsfield Unit (HU) window (lung parenchyma range)
- Denoising via a Laplacian of Gaussian filter (σ = 1.0 mm) to preserve fine textural information
- Cropping around the lung region to reduce computational overhead. Lung masks were extracted from each CT scan using a semi-automated lung segmentation pipeline built on the LUNA16 framework. To concentrate on feature extraction on high-risk parenchymal zones, regions that overlapped with the radiation treatment volume (RTV) were kept.

To enhance ROI localization, planning structures and dose-volume histograms (DVH), if available, were co-registered with the CT scan. Radiomics features were extracted using the PyRadiomics open-source platform [2], which supports IBSI-compliant extraction of interpretable image biomarkers. In total, 120 features per scan were extracted, grouped as follows:

a) First-order statistics (n = 18): Mean, variance, skewness, kurtosis, energy, entropy.

   Reflect intensity distribution and heterogeneity.

b) Shape features (n = 14): Volume, surface area, sphericity, elongation. Describe anatomical distortion due to inflammation.

c) Texture features (n = 88): GLCM (gray-level co-occurrence matrix): correlation, contrast, homogeneity, GLRLM (run-length): long/short run emphasis, GLSZM (size zone): gray level non-uniformity, NGTDM (neighborhood difference): coarseness, strength, GLDM (dependence matrix): dependence non-uniformity.

All features were extracted in 3D using a fixed bin width of 25 HU for gray-level discretization.

2.3   Surrogate Infrared Dataset and Feature Extraction

We use a proxy-based feature fusion approach to mimic functional imaging input because there are no institutional or public datasets that include paired infrared and CT imaging of radiation pneumonitis patients [20]. We use surrogate IR-derived features that were taken from public thermal imaging datasets that deal with respiratory conditions like thoracic inflammation, respiratory instability, and localized hyperthermia that have physiological similarities to early RP.

To train a multimodal classifier, this method combines representative functional descriptors with CT radiomics to create a synthetic but physiologically realistic feature fusion dataset. We used two publicly accessible thermal imaging datasets that record thoracic temperature distributions and respiratory dynamics in patients with acute pulmonary conditions in order to simulate infrared (IR) features pertinent to early radiation-induced pneumonitis (RP) [21, 22]. These datasets offer physiologically significant proxies, reflecting early inflammatory responses, fever, and abnormal ventilation-phenotypes also present in early RP onset-despite the fact that they do not contain RP-specific cases. Three criteria were used to select each dataset: (1) thoracic-level infrared acquisition; (2) related clinical metadata; and (3) public availability for reproducible science. The datasets that were chosen are presented in Table 1, which summarizes their characteristics, resolution, and associated clinical conditions.

Table 1. The selected datasets

Dataset	Description	Resolution	Associated Conditions
PhysioNet COVID-Thermal	Thermal videos of upper body in febrile and afebrile COVID-19 patients	320 × 240	Fever, respiratory distress, asymmetric ventilation
Low-Cost Thermal Pneumonia Pilot	Thermal images and videos of the chest and back in suspected pneumonia cases	320 × 240	Lung inflammation, localized thoracic hyperthermia

By classifying patients into "RP-like" and "non-RP-like" groups according to the clinical annotations that were available (such as fever, tachypnea, COVID-positive, and pneumonia-positive), surrogate labels were established.

These labels were then associated with matched CT radiomics samples to enable feature-level multimodal learning.

The following physiological manifestations observed in early RP are known to be observable via thermal imaging (as validated in non-RP datasets):

- Thoracic hyperthermia due to underlying inflammatory processes
- Left-right temperature asymmetry across the lung fields, indicative of regional dysfunction
- Irregular breathing cycles, captured via frame-wise thermal variance
- Shallow or rapid breathing patterns, indirectly observed via thermal motion dynamics.

These surrogates, although not exclusive to RP, are characteristic of early pulmonary inflammation and serve as functionally plausible indicators in the absence of paired ground-truth RP + IR imaging.

Figure 2 provides a visual comparison between infrared thermography and CT imaging modalities in the thoracic region. The IR image illustrates thermal surface asymmetry and regional hyperthermia, while the CT slice demonstrates underlying anatomical and parenchymal changes. This pairing conceptually supports our fusion rationale, where functionally suggestive IR features are mapped onto structurally informative CT radiomics to simulate an early RP detection scenario. While no actual paired CT-IR dataset of RP exists, this visual juxtaposition conceptually motivates the surrogate alignment proposed in our methodology.

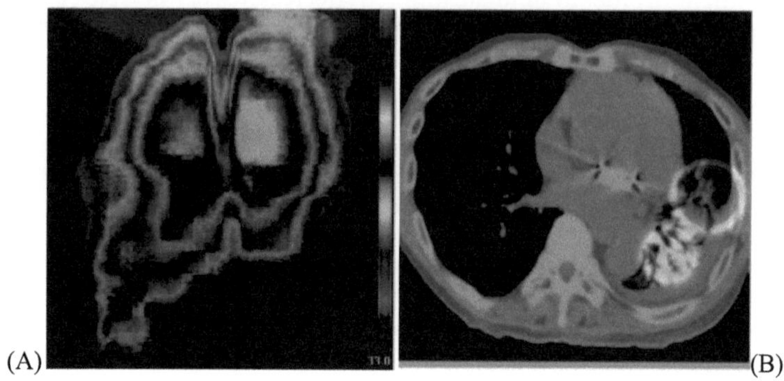

(A)   (B)

Fig. 2. (A) Infrared thermographic image of the chest. (B) CT slice of the lung.

Thermal frames (static or video) were analyzed using a Python-based pipeline that employed:

- ROI Detection: Chest and upper thorax region automatically detected using threshold-based segmentation + pose estimation (OpenPose/MediaPipe).
- Temperature Asymmetry Index:

$$A_{temp} = \left| \mu_{left} - \mu_{right} \right| (\text{in } °C)$$

  where μ indicates average temperature over left/right hemithorax.
- Thermal Respiration Signal Extraction: From time-series video, chest expansion/contraction cycles were derived using frame-wise intensity change.
- Breathing rate estimated via FFT (Fast Fourier Transform) over thermal signal.
- Amplitude variability calculated as:

$$\sigma_{amp} = StdDev(amp)(\text{Breath Peak} - \text{to} - \text{Peak Amplitude})$$

For each IR case, we extracted the following 12 surrogate features, as presented in Table 2

These features were used to characterize functional breathing and thermal behavior, acting as physiological proxies for early RP dynamics.

Table 2. Extracted Thermal Features.

Feature	Description
avg_temp	Average thoracic surface temperature
temp_diff_LR	Mean left-right asymmetry
chest_temp_sd	Temperature standard deviation
breathing_rate	Estimated breaths per minute
breath_amp_mean	Average amplitude of respiration signal
breath_amp_var	Variability in amplitude
thermal_cycle_stability	Periodic coherence of respiration
ROI_area	Size of thermal-emitting chest area
temp_gradient_vert	Vertical thermal gradient
chest_motion_range	Range of motion across frames
IR_entropy	Surface temperature randomness
composite_score	Weighted index combining all above

Each CT case (from Sect. 2.2) was matched with an infrared (IR) thermal feature vector (from Sect. 2.3) through a label-preserving process: CT cases labeled as RP+ were randomly paired with IR profiles from "RP-like" patients (e.g., COVID-19 with fever, pneumonia), while RP-CT cases were matched with IR data from healthy or non-inflammatory subjects, ensuring strict label alignment and no cross-contamination. To avoid overfitting specific samples, multiple random CT-IR pairings were tested, confirming model robustness. Each sample was represented by a fused feature vector combining approximately 100 CT radiomic features with 12 thermal surrogate features, resulting in a total dimensionality of about 112 features per sample.

Each data point was represented as:

$$X_{fused} = [x_{CT1}, x_{CT2}, \ldots, x_{CTn}, x_{IR1}, x_{IR2}, \ldots, x_{IRn}]$$

where x_{CT} are CT radiomic features ($n \approx 100$), and x_{IR} are thermal surrogates ($m = 12$)

To ensure compatibility and prevent modality dominance, all features were standardized via z-score normalization:

$$z_i = \frac{x_i - \mu_i}{\sigma_i}$$

Feature selection was applied prior to model training using: Variance thresholding (min var $= 0.01$), Pearson correlation filtering ($\rho > 0.95$) and Recursive feature elimination (RFE) within cross-validation.

We performed model ablation studies comparing a baseline model that only included CT, an IR-only (functional-only) model, and a fused model that combined CT and IR features to assess whether the surrogate IR features significantly aid in classification. We permuted IR labels and measured the performance drops that resulted in evaluating possible label leakage or overfitting brought on by IR features. Additionally, SHAP analysis was performed post hoc to quantify the importance ranking of IR features within the fused model. We acknowledge that pairing CT and IR samples from different individuals relies on key assumptions: first, that a CT case labeled RP + is physiologically comparable to a thermally abnormal case from another condition (e.g., febrile COVID-19), and second, that surrogate IR features do not introduce bias absent from the CT data.

2.4 Model Architecture and Explainability

We developed a strong supervised classification pipeline to assess the predictive power of the fused multimodal feature set (CT radiomics + surrogate IR features). Transparency and validation are crucial in the medical field, and the chosen machine learning model places a high priority on interpretability, robustness to feature scale, and the ability to handle mixed-type data with complex interactions. After benchmarking several candidate algorithms, including Logistic Regression [23], SVM [24], and XGBoost [25], we decided to use the Random Forest (RF) classifier as the final model [26]. This was brought about by its ability to handle mixed feature types, capture non-linear interactions, perform well on tabular, high-dimensional data, and be compatible with SHAP for explainability. It also naturally resists overfitting through bagging. Table 3 presents the parameters used for the Random Forest classifier from scikit-learn [27].

Table 3. RandomForest Classifier with the following parameters

Parameter	Value	Rationale
n_estimators	300	Stability across trees
max_depth	8	Controls overfitting
min_samples_split	4	Prevents overly specific splits
class_weight	balanced	Accounts for class imbalance
random_state	42	Reproducibility

To reduce over-fitting risk in the 120×112 setting, we used stratified 5-fold cross-validation with all feature-selection and model-tuning confined inside the training folds. We further included a binary IR-dataset indicator as a covariate to mitigate domain shifts across IR repositories. Two robustness controls were performed: (i) label-preserving IR permutation (IR feature vectors randomly permuted within class) to test whether performance relies on physiologically aligned IR signals rather than dataset artefacts; and (ii) Leave-One-IR-Dataset-Out (LODO) evaluation to assess sensitivity to the source of surrogate IR features. Model calibration was examined with reliability curves and the Brier score. Code to reproduce these checks and exact protocols for calibration, permutation, and LODO checks are publicly available in the repository.

3 Results

In this section, we present the quantitative results of the baseline model that solely uses CT-based features versus the suggested multimodal fusion model (CT radiomics + IR thermal features). ROC curves, confusion matrices, and statistical comparisons using accepted evaluation metrics are used to display the results. Standard classification metrics, including accuracy, sensitivity, specificity, precision, F1-score, and AUC-ROC, were used to assess the performance of the proposed CT + IR fusion model and the CT-only baseline model. To guarantee robustness, all metrics were calculated as the average of a 3× repeated 5-fold stratified cross-validation.

3.1 Performance Metrics Overview

The model's overall accuracy rose from 76% to 86%, indicating strong overall performance, despite accuracy's limitations in unbalanced medical datasets. Compared to 68% with CT-only data, the fusion model accurately detects 88% of true RP+ cases. By identifying functional abnormalities like thermal asymmetry before complete radiologic signs manifest, this reduces missed diagnoses and enables earlier intervention. Sensitivity, a crucial metric for early radiation pneumonitis detection, increased significantly from 0.68 to 0.88. A slight improvement in specificity demonstrated that adding infrared features did not increase false positive rates, maintaining model precision. The fusion model's positive predictions for clinical triage or decision support became more dependable as precision increased from 0.77 to 0.85. As demonstrated by the F1-score, which balances recall and precision, the ability to recognize and believe positive RP cases has generally improved, increasing from 0.72 to 0.86. Improved discrimination between RP+ and RP-cases across various decision thresholds was indicated by the area under the ROC curve, which grew from 0.79 to 0.91, for flexible clinical use, this is essential. These gains, particularly in sensitivity (+29%), F1-score (+19%), and AUC (+15%), were consistent across cross-validation folds with low variability, supporting good generalization. Importantly, no trade-offs in specificity or increases in false positives were observed, highlighting the robustness of the fusion approach. In summary, these results support the central hypothesis that early RP detection is improved by combining functional thermal descriptors with anatomical CT features, particularly in cases where CT markers are weak or unclear. Table 4 presents the comparison of model performance (mean ± std).

Beyond discrimination, the fused model showed better probability calibration than the CT-only baseline (lower Brier score; reliability curves provided in the repository). Both robustness controls degraded performance relative to the fused model—IR permutation reduced AUC and sensitivity, and leave-one-IR-dataset-out produced lower but consistent gains over CT-only—supporting that improvements are not driven by surrogate-dataset artefacts. We further analyzed the false negatives (FN) and false positives (FP) to assess the contribution of IR-derived features. False negatives in the CT-only model were correctly classified in the fusion model. This improvement was primarily driven by functional IR features, as these RP+ cases showed subtle or pre-radiographic changes on CT but exhibited increased thermal asymmetry (e.g., temperature difference between left and right sides >1.2 °C) and abnormal respiratory variability (e.g., breath

Table 4. Model Performance Comparison

Metric	CT-only	CT + IR Fusion	Δ (%)
Accuracy	0.76 ± 0.05	0.86 ± 0.04	+13.2%
Sensitivity	0.68 ± 0.06	0.88 ± 0.05	+29.4%
Specificity	0.83 ± 0.04	0.84 ± 0.04	+1.2%
Precision (PPV)	0.77 ± 0.05	0.85 ± 0.03	+10.4%
F1-score	0.72 ± 0.06	0.86 ± 0.04	+19.4%
AUC-ROC	0.79 ± 0.03	0.91 ± 0.02	+15.2%

cycle variation >20%). Such functional patterns allowed the fusion model to correctly reclassify patients that the CT-only model had misclassified as RP-. A subset of false positives in the fusion model was driven by IR-derived indicators, such as elevated chest surface temperature and asymmetry, which may also be present in post-infectious or post-surgical inflammatory states. As a result, these FPs are physiologically explainable rather than genuine model errors.

3.2   Feature Importance and Explainability

To enhance clinical trust and interpretability, we applied SHAP (SHapley Additive Explanations) to both the CT-only and CT + IR fusion models [28]. SHAP allows us to quantify how individual features influence predictions toward or away from RP. In the CT-only model, key features relate to structural and textural lung abnormalities-such as texture complexity and intensity distribution asymmetry-which mainly emerge after RP has progressed, thus limiting early detection capability. When IR-based surrogate features are integrated, the model's important predictors shift notably: surface temperature imbalance between lung regions (*temp_diff_LR*) becomes the most influential, followed by breathing cycle variability and other thermal surface characteristics, reflecting functional changes that precede visible CT alterations. Using a SHAP force plot at the patient level, it was demonstrated that the fusion model accurately predicted RP+ by utilizing surface temperature imbalance and respiratory irregularities identified by IR features, whereas the CT-only model failed to detect early RP signs. According to a cross-modality analysis of SHAP values, CT (52.3%) and IR (47.7%) features contributed equally, with CT features predominating in advanced stages of the disease and IR features providing complementary early diagnostic signals. Across all RP progression stages, this fusion approach yields predictions that are more reliable and comprehensible [29]. All things considered, SHAP demonstrates that the integration of functional thermal markers with anatomical CT radiomics improves early RP detection and provides clear, physiologically significant information to support clinical decision-making. Table 5 presents the top-ranked features contributing to the predictions of the CT + IR fusion model where five stars indicate the highest relative contribution among features and one star the lowest within the top-ranked set.

Table 5. Top-ranked features contributing to the CT + IR fusion model predictions

Rank	Feature Name	Source	SHAP Importance	Interpretation
1	*temp_diff_LR*	IR	★★★★☆	Thermal asymmetry between left/right lung zones
2	*GLCM_Entropy*	CT	★★★★☆	Texture complexity of irradiated lung region
3	*breath_cycle_var*	IR	★★★☆☆	Variability in thermal breathing cycle
4	*FirstOrder_Kurtosis*	CT	★★★☆☆	Localized density irregularity
5	*ir_skewness*	IR	★★☆☆☆	Asymmetry in thermal surface profile

In the SHAP summary plot (as illustrated in Fig. 3), each dot represents a single patient's feature impact on the model's prediction. Red dots indicate high feature values, while blue dots represent low feature values. The horizontal position of the dots shows whether the feature increases (positive SHAP value) or decreases (negative SHAP value) the likelihood of radiation pneumonitis. This color and position coding helps visualize how feature values influence predictions across the cohort.

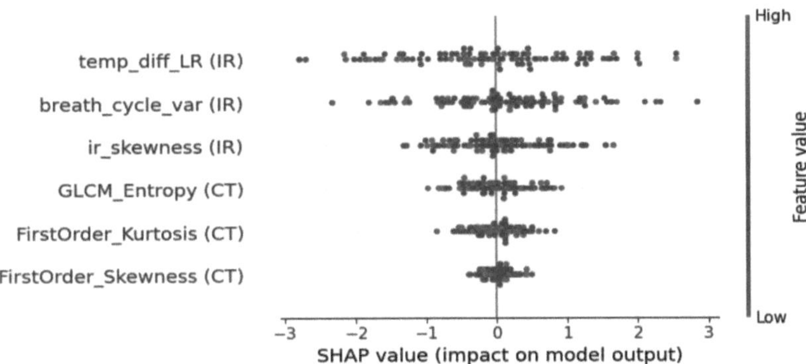

Fig. 3. SHAP Explanation for Model Interpretability

4 Conclusion

To facilitate the early detection of radiation-induced pneumonitis, this study presents the first multimodal AI framework that combines radiomic features from CT scans with functional surrogates derived from infrared thermal imaging. The innovation resides in combining anatomical-textural descriptors taken from high-resolution CT using radiomics with non-invasive surface biomarkers, such as respiratory pattern irregularities and thermal asymmetry. The resulting machine learning model exhibits significantly better sensitivity (+29.4%) and AUC-ROC (+15.2%) compared to CT-only baselines, particularly in cases of early or borderline RP. In line with the physiological recognition that functional disruptions often occur prior to radiologic abnormalities, SHAP analysis indicates that IR features are the most important factor in early predictions. Clinically, this fusion approach provides an interpretable, low-cost early-warning tool that can be deployed in radiation oncology follow-up clinics. Unlike "black box" deep learning models, our method emphasizes transparency and interpretability, in line with trustworthy AI principles. Compared to existing literature that either relies solely on radiomics or applies IR imaging in unrelated pulmonary settings, this work is the first to fuse both modalities for RP, providing clinically relevant, interpretable, and quantifiable results. The use of surrogate infrared data from non-RP sources (like COVID-19 datasets), the lack of paired ground-truth thermographic recordings, the small sample size, and the absence of real-world environmental factors are some of the limitations. The main limitation is the synthetic multimodal pairing: CT radiomics from RP-labeled patients were fused with surrogate IR features from different individuals and conditions (e.g., COVID-19, pneumonia), introducing cross-patient/cross-disease pairing, possible temporal misalignment, and population heterogeneity. We mitigated these via harmonization (z-scoring and an IR-dataset indicator covariate), label-preserving IR permutation, leave-one-IR-dataset-out evaluation, and probability calibration; still, we deliberately avoid generalization claims. The contribution is methodological and feasibility-oriented: functionally motivated IR descriptors provide complementary signals to CT radiomics for very-early RP detection. Future work will prospectively acquire paired CT–IR thoracic data post-RT to validate these findings clinically. Future directions include developing attention-based hybrid models, integrating real-time IR monitoring into digital twin frameworks, and creating prospectively acquired CT + IR RP datasets. Overall, this study offers evidence that functional thermal descriptors can improve early RP detection and lay the groundwork for future clinical implementation when carefully matched with anatomical data.

Acknowledgments. This study used exclusively publicly available datasets that were anonymized and collected in accordance with the ethical standards of the respective institutions. No new patient data were collected, and no identifiable personal information was processed. As such, ethics approval and informed consent were not required for the purposes of this research. The authors would like to acknowledge the providers and curators of open-access imaging databases, including thermal datasets used for surrogate modeling. This work was conducted independently and did not receive any specific funding. However, the research benefited from academic guidance, prior published literature, and collaborative discussions within the scientific community.

Disclosure of Interests. The authors have no competing interests to declare that are relevant to the content of this article.

Data and Code Availability Statement. Code for reproducing the experiments, including preprocessing, training, and evaluation pipelines, are available at: https://github.com/sotraptis/multimodal-rp-detection.

References

1. Bledsoe, T.J., Nath, S.K., Decker, R.H.: Radiation pneumonitis. Clin. Chest Med. **38**(2), 201–208 (2017). https://doi.org/10.1016/j.ccm.2016.12.004
2. Yue, J., et al.: Patient-reported lung symptoms as an early signal of impending radiation pneumonitis in patients with non-small cell lung cancer treated with chemoradiation: an observational study. Qual. Life Res. **27**(6), 1563–1570 (2018). https://doi.org/10.1007/s11 136-018-1834-3
3. Raptis, S., Ilioudis, C., Theodorou, K.: Uncovering the diagnostic power of radiomic feature significance in automated lung cancer detection: an integrative analysis of texture, shape, and intensity contributions. BioMedInformatics **4**(4), 2400–2425 (2024). https://doi.org/10.3390/biomedinformatics4040129
4. Thomas, R., Chen, Y.-H., Hatabu, H., Mak, R.H., Nishino, M.: Radiographic patterns of symptomatic radiation pneumonitis in lung cancer patients: imaging predictors for clinical severity and outcome. Lung Cancer **145**, 132–139 (2020). https://doi.org/10.1016/j.lungcan.2020.03.023
5. Lahiri, B.B., Bagavathiappan, S., Jayakumar, T., Philip, J.: Medical applications of infrared thermography: a review. Infrared Phys. Technol. **55**(4), 221–235 (2012). https://doi.org/10.1016/j.infrared.2012.03.007
6. Perpetuini, D., Filippini, C., Cardone, D., Merla, A.: An overview of thermal infrared imaging-based screenings during pandemic emergencies. Int. J. Environ. Res. Public Health **18**(6), 3286 (2021). https://doi.org/10.3390/ijerph18063286
7. Kakileti, S.T., Dalmia, A., Manjunath, G.: Exploring deep learning networks for tumour segmentation in infrared images. Quant. InfraRed Thermogr. J. **17**(3), 153–168 (2020). https://doi.org/10.1080/17686733.2019.1619355
8. Etehadtavakol, M., Ng, E.Y.K.: Survey of numerical bioheat transfer modelling for accurate skin surface measurements. Therm. Sci. Eng. Prog. **20**, 100681 (2020). https://doi.org/10.1016/j.tsep.2020.100681
9. Mukhmetov, O., Zhao, Y., Mashekova, A., Zarikas, V., Ng, E.Y.K., Aidossov, N.: Physics-informed neural network for fast prediction of temperature distributions in cancerous breasts as a potential efficient portable AI-based diagnostic tool. Comput. Methods Programs Biomed. **242**, 107834 (2023). https://doi.org/10.1016/j.cmpb.2023.107834
10. Gangadharan, C., et al.: Artificial intelligence enhanced thermal breast imaging in the diagnosis of invasive breast cancer: a study of 2 case reports. Eur. J. Med. Case Rep. **7**(2) (2023). https://doi.org/10.24911/ejmcr/173-1655212916
11. Raptis, S., Softa, V., Angelidis, G., Ilioudis, C., Theodorou, K.: Automation radiomics in predicting radiation pneumonitis (RP). Automation **4**(3), 191–209 (2023). https://doi.org/10.3390/automation4030012
12. Lipkova, J., et al.: Artificial intelligence for multimodal data integration in oncology. Cancer Cell **40**(10), 1095–1110 (2022). https://doi.org/10.1016/j.ccell.2022.09.012
13. Kakileti, S.T., Manjunath, G., Schwartz, R.G., Ng, E.Y.-K.: Artificial Intelligence Over Infrared Images for Medical Applications: Third International Conference, AIIIMA 2024, Virtual Event, November 9, 2024: Proceedings. Lecture Notes in Computer Science, no. 15279. Springer, Cham (2025)
14. Van Griethuysen, J.J.M., et al.: Computational radiomics system to decode the radiographic phenotype. Cancer Res. **77**(21), e104–e107 (2017). https://doi.org/10.1158/0008-5472.can-17-0339
15. Nohara, Y., Matsumoto, K., Soejima, H., Nakashima, N.: Explanation of machine learning models using shapley additive explanation and application for real data in hospital. Comput. Methods Programs Biomed. **214**, 106584 (2022). https://doi.org/10.1016/j.cmpb.2021.106584

16. Carrington, A.M., et al.: Deep ROC analysis and AUC as balanced average accuracy, for improved classifier selection, audit and explanation. IEEE Trans. Pattern Anal. Mach. Intell. **45**(1), 329–341 (2022)

17. Wee, L., Aerts, H.J., Kalendralis, P., Dekker, A.: Data from NSCLC-radiomics-interobserver1. Cancer Imaging Arch. (2019). https://doi.org/10.7937/TCIA.2019.CWVLPD26

18. Wee, L., Aerts, H., Kalendralis, P., Dekker, A.: RIDER lung CT segmentation labels from: decoding tumour phenotype by noninvasive imaging using a quantitative radiomics approach. Cancer Imaging Arch. (2020). https://doi.org/10.7937/TCIA.2020.JIT9GRK8

19. Aerts, H.J.W.L., et al.: Data from NSCLC-radiomics-genomics. Cancer Imaging Arch. (2015). https://doi.org/10.7937/K9/TCIA.2015.L4FRET6Z

20. Gao, Z., Zhang, Y., Li, Y.: Extracting features from infrared images using convolutional neural networks and transfer learning. Infrared Phys. Technol. **105**, 103237 (2020). https://doi.org/10.1016/j.infrared.2020.103237

21. Tamez-Peña, J., Yala, A., Cardona, S., Ortiz-Lopez, R., Trevino, V.: Upper body thermal images and associated clinical data from a pilot cohort study of COVID-19. PhysioNet. https://doi.org/10.13026/WFR2-5973

22. Qu, Y., Meng, Y., Fan, H., Xu, R.X.: Low-cost thermal imaging with machine learning for non-invasive diagnosis and therapeutic monitoring of pneumonia. Infrared Phys. Technol. **123**, 104201 (2022). https://doi.org/10.1016/j.infrared.2022.104201

23. Boateng, E.Y., Abaye, D.A.: A review of the logistic regression model with emphasis on medical research. J. Data Anal. Inf. Process. **07**(04), 190–207 (2019). https://doi.org/10.4236/jdaip.2019.74012

24. Peker, M.: A decision support system to improve medical diagnosis using a combination of k-medoids clustering based attribute weighting and SVM. J. Med. Syst. **40**(5) (2016). https://doi.org/10.1007/s10916-016-0477-6

25. Zhang, X., Yan, C., Gao, C., Malin, B.A., Chen, Y.: Predicting missing values in medical data via XGBoost regression. J. Healthc. Inform. Res. **4**(4), 383–394 (2020). https://doi.org/10.1007/s41666-020-00077-1

26. Yang, F., Wang, H., Mi, H., Lin, C., Cai, W.: Using random forest for reliable classification and cost-sensitive learning for medical diagnosis. BMC Bioinform. **10**(S1) (2009). https://doi.org/10.1186/1471-2105-10-s1-s22

27. Pedregosa, F., et al.: Scikit-learn: machine learning in Python. J. Mach. Learn. Res. **12**, 2825–2830 (2011)

28. Raptis, S., Ilioudis, C., Theodorou, K.: From pixels to prognosis: unveiling radiomics models with SHAP and LIME for enhanced interpretability. Biomed. Phys. Eng. Express **10**(3), 035016 (2024). https://doi.org/10.1088/2057-1976/ad34db

29. Jain, V., Berman, A.: Radiation pneumonitis: old problem, new tricks. Cancers **10**(7), 222 (2018). https://doi.org/10.3390/cancers10070222

Customized CNN Based Multiclass Classification of Childhood Obesity Using Infrared Thermal Imaging

Richa Rashmi[1,2] and U. Snekhalatha[1(✉)]

[1] Department of Biomedical Engineering, College of Engineering and Technology, SRM Institute of Science and Technology, Kattankulathur, Chennai 603203, Tamil Nadu, India
snehalau@srmist.edu.in

[2] Quality Assurance Cell, ANIIMS and GB Pant Hospital, Sri Vijaya Puram 744103, Andaman and Nicobar Island, India

Abstract. Childhood obesity is an escalating global health concern, emphasizing the need for early and effective detection methods. Traditional assessment techniques for fat distribution are often invasive and unsuitable for pediatric populations. This study explores a non-invasive alternative by integrating infrared thermal imaging (IRT) with deep learning models for classifying obesity levels among children. Thermal images were captured from the abdominal and neck regions of 150 children categorized as normal, overweight, or obese. A patch-based image analysis was performed using three models: a custom convolutional neural network (CNN), DenseNet121, and Inception ResNetV2. The custom CNN, developed and trained from scratch, consistently outperformed the other models, achieving accuracy rates of 93.1% for the abdomen and 91.6% for the neck, with balanced classification across all BMI categories. The abdominal region showed better discriminatory patterns, supported by higher AUC scores. An ablation study further validated the network's robustness and highlighted the influence of hyperparameter optimization. The results indicate that deep learning applied to thermal imaging is a promising, non-invasive tool for early obesity screening in children.

Keywords: Childhood Obesity · Thermal Imaging · Convolutional Neural Network · Deep Learning

1 Introduction

Childhood obesity is a pressing global health issue with long-term risks such as diabetes, hypertension, and cardiovascular complications. Traditionally, obesity assessment has relied on anthropometric indices such as body mass index (BMI) and waist-to-hip ratio. However, these measures are limited in pediatric populations due to growth-related variability and poor discrimination of adiposity

S. T. Kakileti et al. (Eds.): AIIIMA 2025, LNCS 16308, pp. 95–112, 2026.
https://doi.org/10.1007/978-3-032-10990-3_7

distribution [1]. Advances in imaging techniques have provided alternatives for more objective evaluation. Infrared thermal imaging has emerged as a promising, non-invasive, radiation-free modality capable of capturing surface temperature variations that reflect underlying metabolic activity. Earlier studies have applied thermal imaging for fever screening, breast cancer detection, and vascular diagnostics, demonstrating its clinical reliability [2–4]. Its application to obesity assessment remains comparatively limited. Rashmi et al. (2021) presented one of the earliest efforts, using thermograms from different body regions combined with machine learning classifiers to differentiate obese from non-obese children [5]. More recent work has refined this approach by proposing a lightweight CNN trained on thermal images for obesity early detection, achieving comparable performance to deeper networks such as DenseNet and MobileNet while reducing computational cost [6].

Parallel to these developments, deep learning models have shown significant promise in health-related imaging tasks. Convolutional neural networks (CNNs) have shown exceptional performance in medical imaging applications such as classification, detection, and segmentation [7]. Furthermore, a scoping review highlighted the growing role of AI in obesity research, while also pointing out that imaging-based deep learning studies remain sparse compared to anthropometry-based approaches [8]. Umapathy et al. [9] demonstrated that CNNs combined with thermal imaging could effectively classify adult subjects as obese or normal based on skin surface temperature from body regions such as the abdomen and thigh. Despite these advances, there is still limited literature on thermal imaging for multiclass obesity classification in children, particularly with end-to-end deep learning architectures.

This gap motivates our work, where we propose a multiclass classification framework for childhood obesity. It evaluates thermal images from the abdomen and neck using Modified ResNetV2, DenseNet121, and a custom-designed CNN. An ablation study is also conducted to examine how optimizers and learning rates influence model performance.

2   Methodology

2.1   Data Collection

The present study received ethical approval from the SRM Institutional Ethics Committee (IEC No: 17401IEC/2019). A total of 150 children, aged between 7 and 14 years, were enrolled from local schools in and around Kattankulathur, Chennai, India. Participants were evenly divided into three groups according to body mass index (BMI) classifications according to World Health Organization (WHO) guidelines: 50 were classified as normal (BMI 85th percentile), 50 as overweight (BMI between 85 and 95 percentile) and 50 as obese (BMI < 95 percentile). Equal representation of boys and girls was maintained across all groups. Children who were unwell, on medication, or had a fever at the time of data collection were excluded from the study. The purpose and procedures of the research were clearly explained to the school authorities and parents or guardians. Written informed consent was obtained from the guardians, and verbal assent was taken from the children. All collected data were handled confidentially, and participant anonymity was maintained throughout the study.

2.2   Baseline Measurements and Experimental Setup

Before image acquisition, participants were acclimatized in a temperature-controlled room maintained at approximately 20°C to ensure consistent thermal conditions. Each child changed into lightweight cotton clothing to minimize insulation effects. Baseline anthropometric measurements namely height and weight were recorded using a calibrated stadiometer and digital weighing scale, respectively, and used to calculate BMI for classification into normal, overweight, or obese groups. Thermal imaging was performed using a FLIR A325SC infrared camera mounted on a fixed tripod. Participants stood upright at a standardized 600 mm distance, marked on the floor, with arms slightly raised from the torso to ensure clear visibility of the abdomen and neck regions. Individual thermal images were captured for each region to facilitate anatomical comparison.

2.3   Skin Surface Temperature Measurement

Thermal images were captured for all 150 participants across two study regions. Imaging was performed using a high-resolution FLIR A325SC thermal camera (FLIR Systems, Inc., Wilsonville, Oregon, USA), equipped with a 640×512 pixel uncooled microbolometer sensor. This system allowed for the detection of fine variations in skin surface temperature across the selected regions. Each thermogram was processed and analyzed using FLIR Tools software, which facilitated accurate temperature mapping and data extraction. To ensure consistency in temperature measurements, a circular Region of Interest (ROI) was manually placed on each image within anatomically consistent locations for all subjects. The diameter and placement of the ROI were standardized and kept constant across all participants and regions. This controlled approach ensured reliable comparison of thermal patterns among normal, overweight and obese children for each anatomical site.

2.4   Patch-Based Processing

Patch-based processing is widely used in medical image analysis to enhance classification and segmentation performance. By dividing large images into smaller regions, models can capture fine-grained local features that might be overlooked in full-scale analysis [10]. In this study, thermal images from the abdomen, neck, and gluteal regions were automatically segmented into ten non-overlapping patches of size 128×128 pixels using a predefined grid-based algorithm implemented in the deep learning preprocessing code. This ensured that patch extraction was consistent, reproducible, and free from manual intervention or bias.

The technique increased data variability, emphasized region-specific thermal patterns, and improved the model's ability to distinguish between normal, overweight, and obese categories. Additionally, patch-based input helped mitigate overfitting by enriching the training dataset without synthetic augmentation.

2.5 Deep Learning Architectures

Deep learning models extract features automatically by passing input data through multiple layers that learn hierarchical representations. This makes them well-suited for medical image classification [11]. In this study, three CNN architectures were used to classify childhood obesity using thermal image patches: ResNetV2, DenseNet121, and a custom-built model named Custom-Net. ResNetV2 uses skip connections to improve training in deeper networks, while DenseNet121 introduces dense inter-layer connections to enhance feature propagation and reduce gradient loss. CustomNet was specifically designed for this application as a lightweight model optimized for thermal images with lower computational demand. All three networks were trained to distinguish between normal, overweight, and obese classes based on thermal data from key anatomical regions.

Transfer Learning. Transfer learning is a widely adopted approach in deep learning where pre-trained models, originally developed for large-scale datasets, are adapted to new, domain-specific tasks. This technique significantly reduces training time and allows for improved model performance, especially when working with relatively smaller datasets, such as medical or thermal images [12]. In the current study, transfer learning was applied to the ResNetV2 and DenseNet121 architectures. The initial layers of these models are responsible for capturing low-level features such as edges and textures were retained from pre-training on the ImageNet dataset, while the final classification layers were reconfigured to suit the specific task of classifying thermal images into normal, overweight, and obese categories. The modified section of the network included a Global Average Pooling (GAP) layer followed by two fully connected layers and a softmax activation layer for multi-class output. A comparatively low learning rate (0.0001) was used during training to prevent significant updates to the pre-trained weights, thereby preserving the general features learned from the base dataset. This approach helped reduce overfitting, ensured stable convergence, and improved the generalization performance of the models on thermal image data. The CustomNet architecture, being developed from scratch, was trained entirely on the dataset without the use of pre-trained weights.

Enhanced Inception ResNet V2. Inception ResNet-v2 is a deep convolutional neural network that merges concepts from the Inception architecture and ResNet. Introduced by Google, it features Inception modules for multi-scale feature extraction and residual connections for improved training of deep networks [13]. The network starts by getting the input images of size $128 \times 128 \times 3$ which is

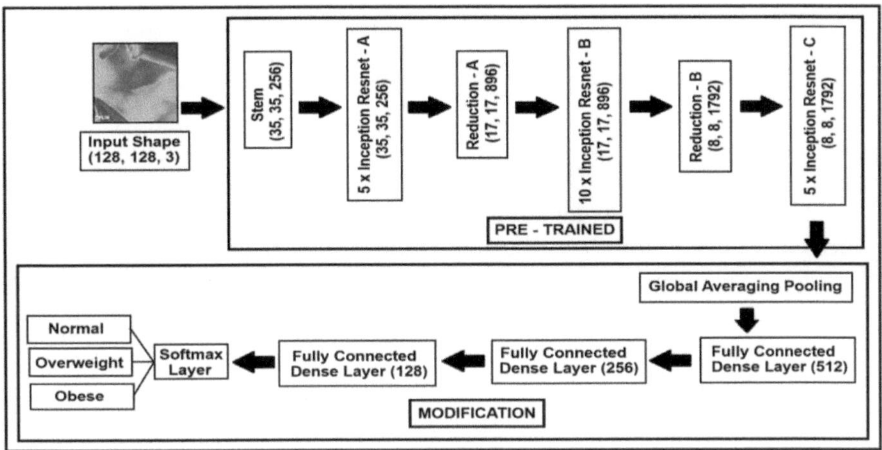

Fig. 1. Architecture diagram of enhanced Inception ResNet V2.

fed into a stem block to process the incoming input images as depicted in Fig. 1. Then the feature maps are made to pass through the inception layers which provides a robust training environment to the incoming feature maps by aiding in gradient flow during backpropagation process. Furthermore, the enhanced feature maps are fed into the reduction blocks which works on the combination of convolution and pooling operation.

This process aids in diminish the spatial dimensions of the incoming feature maps. Then the global average pooling layer (GAP) is interspersed with the Inception-ResNet module hierarchy. A concise representation is produced by this layer as it calculates the mean value for every channel in the feature maps. Moreover, three fully connected dense layers constitute the last layers of Inception ResNet-v2. As a general practice, class probabilities are generated utilizing a softmax activation function. In order to enable the network to generate precise predictions, this last layer converts the high-level features into a probability distribution across all potential classifications to classify normal, overweight and obese children.

Enhanced DenseNet 121. DenseNet-121 is a CNN architecture that is noteworthy for its novel implementation of dense blocks [14]. One distinguishing attribute of these blocks is that every layer within the same block acquires direct input from every preceding layer [15]. The input image of size $128 \times 128 \times 3$ is fed into the convolutional layer which helps in detecting the features in them. This layer has a pooling operation to reduce the spatial dimensions of the feature maps. Then the feature maps are fed into the dense block that promotes feature reuse, facilitates gradient flow during training and enhances parameter efficiency by concatenating the feature maps from all previous layers. The transition block then receives the incoming feature maps from the dense layer as depicted in Fig. 2. The transition layer helps in reducing spatial dimensions and channel numbers in the incoming feature maps. This layer facilitates compression and down-sampling in order to prepare the network for the next dense block by managing computational resources and contributing to overall efficiency of the DenseNet 121 network. In order to decrease the spatial dimensions of the feature maps, GAP layer is implemented which helps in calculating the mean value for every channel thereby providing enhanced generalization. Then the three fully connected layers are introduced in the last stage prior to classification with 512, 256 and 128 channels. Finally, a softmax activation function is utilized to convert the output of the network into different probabilities of classes (normal, overweight and obese).

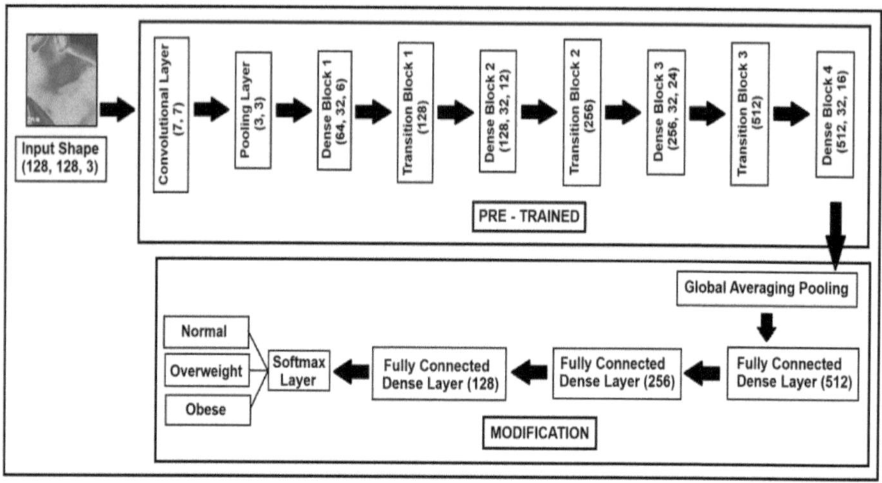

Fig. 2. Architecture diagram of enhanced DenseNet 121.

Customized CNN. A customized deep learning network was developed with the specific purpose of classifying children as normal, overweight and obese based on regional thermograms. The thermograms of size $128 \times 128 \times 3$ are fed as an input image into the first 2D convolutional layer. Three convolutional blocks were employed, each containing 32, 64, and 128 convolution filters, respectively, alongside ReLU activation functions in every layer. By utilizing these eight convolutional layers, complex features were extracted from the original images while the spatial relationships between pixels were maintained. Following that, the inputs were downsampled using a (2,2) max-pooling layer, which concentrated on the most prominent features within the region of the feature maps produced by the convolutional layers. The feature vector that served as the input for the subsequent dense layer was generated through the transformation of the resulting feature maps into a 1D array using the flatten layer, which received the resultant output from the final max-pooling layer. The customised network consisted of three dense layers, each of which was trained using the sigmoid activation function: DL1 (1024 channels), DL2 (128 channels), and DL3 (64 channels) as depicted in Fig. 3. Subsequently, the softmax layer was used to classify children as normal, overweight, or obese. Although the dataset consisted of 150 subjects, the patch-based strategy expanded the dataset to 3,000 thermal patches across anatomical regions, thereby providing sufficient training samples for model learning. The Custom CNN was intentionally designed as a lightweight architecture with fewer trainable parameters compared to conventional deep networks. This design choice ensured that the model could be trained effectively on a relatively small dataset while reducing the risk of overfitting and maintaining stable convergence.

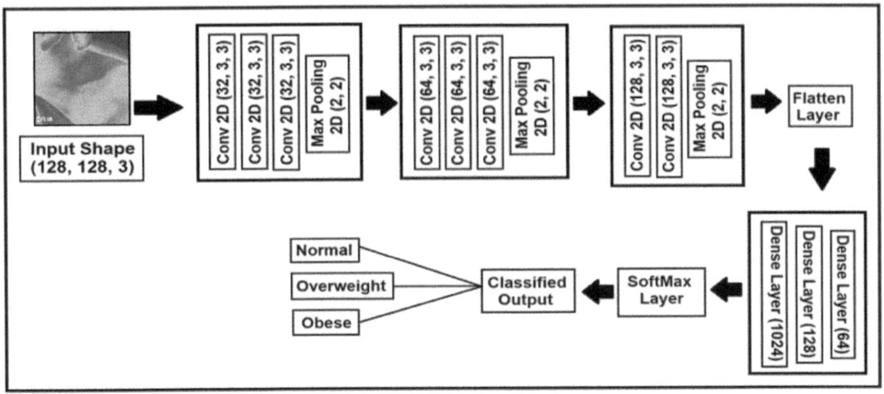

Fig. 3. Architecture of the Customized CNN Network.

2.6   Training and Testing

For this study, thermal images were acquired from two anatomical regions namely abdomen and neck region for a total of 150 children (50 normal, 50 overweight, and 50 obese). From each region, ten non-overlapping patches of size 128×128 pixels were extracted per image, resulting in a comprehensive dataset of 3,000 patches (150 subjects $\times$ 2 regions $\times$ 10 patches). To develop and evaluate the deep learning models, the dataset was divided into training, validation, and testing subsets. For each anatomical region, 20 thermal images were set aside exclusively for testing to ensure unbiased evaluation. The remaining 130 images per region were used for training and validation. This yielded 2,600 patches per region for model development, which were randomly split into 70% training and 30% validation sets. All models were trained separately for each anatomical region using their respective patch datasets. A total of 10 training epochs were executed for each network architecture—Modified ResNetV2, Modified DenseNet121, and CustomNet—using categorical cross-entropy as the loss function and softmax activation for multiclass classification. Model performance

was evaluated on the reserved test set of 600 patches per region (20 images ×
10 patches × 3 classes), and classification metrics were computed for normal,
overweight, and obese categories. This standardized training and testing protocol ensured consistent assessment of model performance across both anatomical
regions.

2.7   Ablation Study

In the context of CNN-based applications, conducting an ablation study is a
standard practice to assess the model's stability and performance when various
layers and hyperparameters are modified or removed [16]. Such modifications can
result in unchanged, enhanced or diminished performance. Typically, optimizing hyperparameters like the optimizer and learning rate can lead to improved
accuracy. Alterations to the model's architecture also significantly affect overall
performance [17]. In this study, we systematically explored different configurations of the proposed Customized CNN model by selectively altering components
and parameters. This investigation included two case studies, with findings thoroughly analysed for each study region. The results indicate an increase in overall
accuracy, demonstrating the efficacy of this approach for this study. Detailed
results of this ablation study are presented in Sect. 3.3.

2.8   Statistical Analysis

The data were analyzed using SPSS software version 19.0, and a one-way
ANOVA test was conducted to determine whether significant differences in
regional skin surface temperatures existed among the normal, overweight, and
obese groups across the abdomen, neck, and gluteal regions.

3   Results

3.1   Regional Temperature Analysis

The average skin surface temperatures at the abdomen and neck regions were
evaluated across three BMI-based groups: normal, overweight, and obese. Temperature values were extracted from fixed circular regions of interest (ROI) on
the thermal images, as shown in Fig. 4. For the abdominal region, the mean temperature recorded for normal participants was $36.9 \pm 0.9\,°C$, which decreased to
$33.8 \pm 1.0\,°C$ in overweight and further to $33.1 \pm 1.1\,°C$ in obese individuals. The
largest temperature reduction of 10.3% was observed between the normal and
obese groups, with a statistically significant difference ($p = 0.0002$). Comparisons
between normal vs overweight and overweight vs obese groups also revealed significant differences, with 8.4% and 2.07% temperature reductions, respectively
($p < 0.05$ for both). In the neck region, a similar trend was observed. Normal
participants exhibited a mean surface temperature of $37.2 \pm 0.6\,°C$, compared
to $34.7 \pm 0.7\,°C$ in overweight and $33.9 \pm 1.0\,°C$ in obese children. The overall

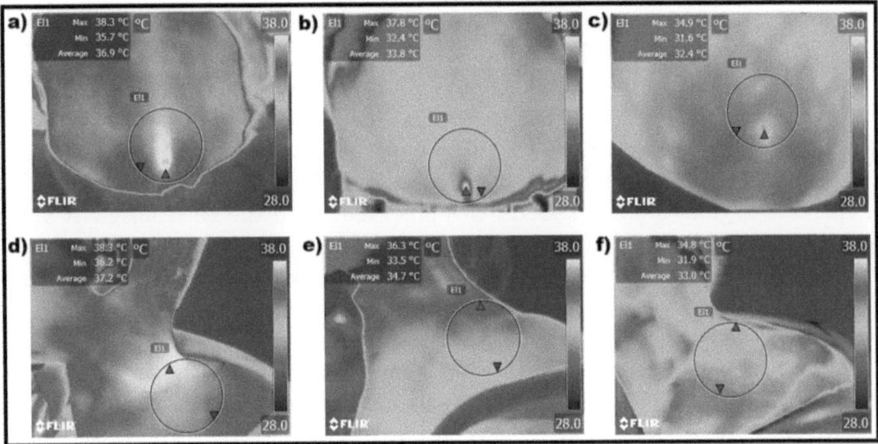

Fig. 4. Thermal images showing region-specific skin surface temperature distribution across BMI categories. (a–c) Abdominal thermograms for (a) normal, (b) overweight, and (c) obese children respectively, with clearly marked circular Regions of Interest (ROIs) used for average temperature extraction. (d–f) Neck region thermograms for (d) normal, (e) overweight, and (f) obese children, showing a progressive decline in surface temperature with increasing BMI.

Table 1. Comparison of regional skin surface temperature values, percentage differences, and p-values across BMI categories for abdomen and neck regions.

Study Region	Normal (°C)	Overweight (°C)	Obese (°C)	Normal vs Overweight		Overweight vs Obese		Normal vs Obese	
				% Diff.	p-value	% Diff.	p-value	% Diff.	p-value
Abdomen	36.9 ± 0.9	33.8 ± 1.0	33.1 ± 1.1	8.40%	0.0004	2.07%	0.0021	10.30%	0.0002
Neck	37.2 ± 0.6	34.7 ± 0.7	33.9 ± 1.0	6.72%	0.0006	2.31%	0.0035	8.86%	0.0003

temperature drop between normal and obese groups was 8.86%, again statistically significant ($p = 0.0003$). Temperature reductions between normal and overweight (6.72%) and between overweight and obese (2.31%) were also significant ($p < 0.05$) as depicted in Table 1. These findings confirm a consistent and significant decline in skin surface temperature with increasing body mass index across both abdominal and neck regions.

3.2   Deep Learning Performance Analysis

To determine which anatomical region yields better classification accuracy for pediatric obesity using thermal imaging, deep learning models were trained and evaluated independently on thermograms of the abdomen and neck. Three models were investigated: ResNetV2, DenseNet121, and a Custom CNN. Transfer learning was utilized for ResNetV2 and DenseNet121, retaining the feature extraction layers from their ImageNet training and fine-tuning the classification layers for the current task. The Custom CNN, developed specifically for this study, was trained from scratch on the collected thermal images. For the abdominal region, all models demonstrated strong classification performance. The Custom CNN achieved the highest accuracy at 93.1% with an AUC of 0.96, followed by DenseNet121 with 90.8% accuracy and ResNetV2 with 87.6% accuracy as depicted in Table 2. Additionally, the Custom CNN maintained balanced precision, recall, and F1-scores exceeding 0.90 across all BMI categories. In the neck region, a similar trend was observed, with the Custom CNN again outperforming the other models, achieving 91.6% accuracy and an AUC of 0.93. DenseNet121 attained 88.6% accuracy, while ResNetV2 reached 86.6% accuracy. While classification performance for the neck was slightly lower compared to the abdomen, the Custom CNN showed consistent results across both regions, indicating robust generalization to different anatomical sites. Further examination of precision, recall, and F1-score metrics showed that all models performed best in detecting obese participants, especially in the neck region where ResNetV2 reached 100% recall for the obese category. In contrast, classification of the overweight group was more challenging, with ResNetV2 showing the lowest F1 score for this category in the neck region, suggesting overlap between overweight and other BMI classes.

To assess model behavior during training, accuracy curves over 30 epochs were plotted for ResNetV2, DenseNet121, and a Customized CNN, using thermograms from both the abdomen and neck regions as depicted in Fig. 5 . The Customized CNN showed stable and consistent training with minimal overfitting, particularly for the abdominal region. In contrast, DenseNet121 and ResNetV2 displayed more fluctuations, especially on neck data, indicating higher sensitivity to thermal variation. Nevertheless, all models achieved high validation accuracy by the final epochs, consistent with their classification metrics. ROC curves for the best-performing model (Customized CNN) showed excellent discrimination. For the abdomen region, the model achieved a micro-average AUC of 0.96 and macro-average AUC of 0.95 as shown in Fig. 6. The Customized CNN consistently demonstrated high sensitivity and specificity across classes, particularly in the abdominal region, reinforcing its robustness for childhood obesity classification as depicted in Table 3. Overall, the Customized CNN outperformed other models, with the abdomen showing clearer thermal patterns than the neck. ROC analysis confirmed its reliability, supporting abdominal thermal imaging as an effective noninvasive tool for classifying childhood obesity.

3.3   Results of Ablation Study

As part of the ablation study, four experiments were conducted by modifying key components of the Customized CNN to improve classification accuracy. The focus was on evaluating different optimizers and learning rates. Results, summarized in Table 4, showed that the Adam optimizer with a 0.001 learning rate achieved the best performance, yielding test accuracies of 93.4% for abdomen and 91.8% for neck regions along with the lowest corresponding test losses of 0.07 and 0.08 respectively. These results highlight the importance of optimizer selec-

Table 2. Classification performance of different models for abdomen and neck regions.

Model	Class	Precision	Recall	F1-Score	Accuracy	AUC
Abdomen						
Inception ResNet V2	Normal	0.86	0.97	0.91	87.6%	0.90
	Overweight	0.84	0.79	0.82		
	Obese	0.93	0.86	0.89		
	Average	0.88	0.87	0.87		
DenseNet 121	Normal	0.96	0.92	0.94	90.8%	0.91
	Overweight	0.82	0.96	0.88		
	Obese	0.98	0.84	0.91		
	Average	0.92	0.91	0.91		
Customized CNN	Normal	0.95	0.95	0.95	93.1%	0.96
	Overweight	0.92	0.89	0.90		
	Obese	0.92	0.95	0.94		
	Average	0.93	0.93	0.93		
Neck						
Inception ResNet V2	Normal	0.61	0.78	0.67	86.6%	0.90
	Overweight	0.78	0.67	0.58		
	Obese	0.83	1.00	0.91		
	Average	0.88	0.87	0.85		
DenseNet 121	Normal	0.86	0.97	0.91	88.6%	0.90
	Overweight	0.84	0.79	0.82		
	Obese	0.93	0.86	0.89		
	Average	0.91	0.88	0.87		
Customized CNN	Normal	0.92	0.94	0.93	91.6%	0.93
	Overweight	0.89	0.91	0.90		
	Obese	0.94	0.89	0.91		
	Average	0.92	0.92	0.92		

tion and learning rate tuning in enhancing model performance, offering broader insights into deep learning optimization strategies.

The ablation study (Table 5) highlights the contribution of individual architectural components of the proposed Custom CNN. The baseline model, which included convolution, batch normalization, dropout, and dense layers, achieved the best overall performance, confirming the importance of a balanced architecture. Removing dropout or batch normalization led to a clear decline in accuracy and F1-score, indicating issues of overfitting and unstable training. Similarly, reducing dense units or convolutional depth lowered feature discrimination, though the performance drop was less severe. These findings emphasize

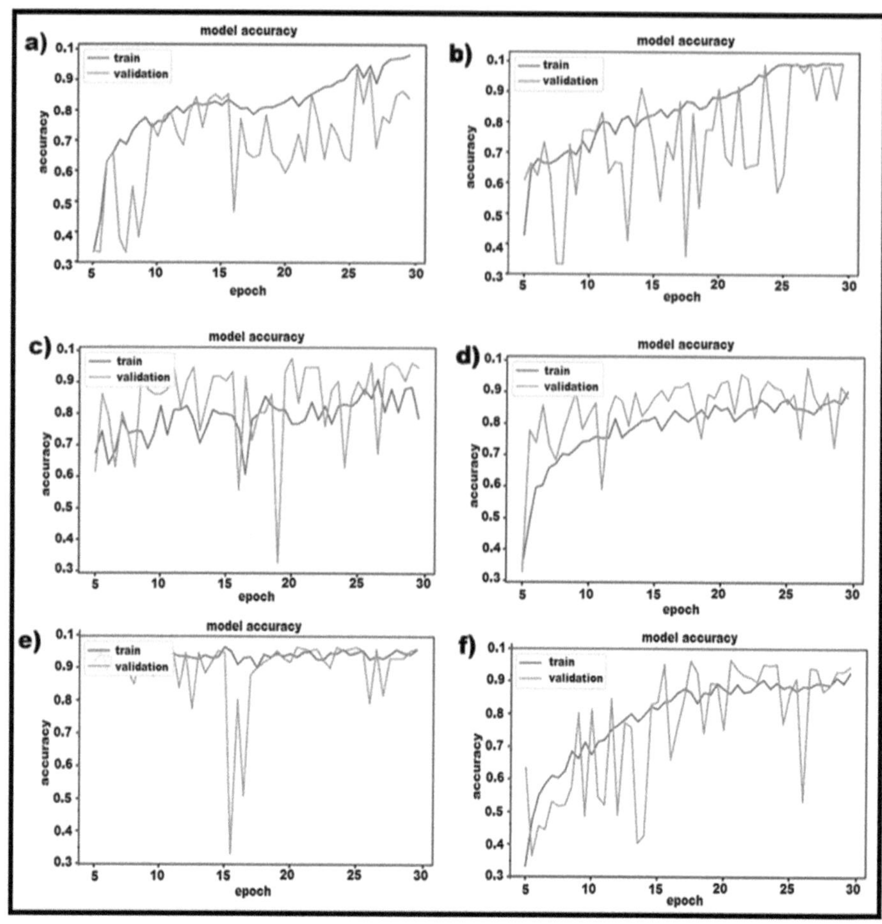

Fig. 5. Model accuracy curves over 30 epochs showing training and validation performance of Inception ResNetV2 (a–b), DenseNet121 (c–d), and Customized CNN (e–f) for thermal images from the abdomen (a, c, e) and neck (b, d, f) regions.

that each component plays a complementary role in ensuring the robustness and generalization of the proposed network.

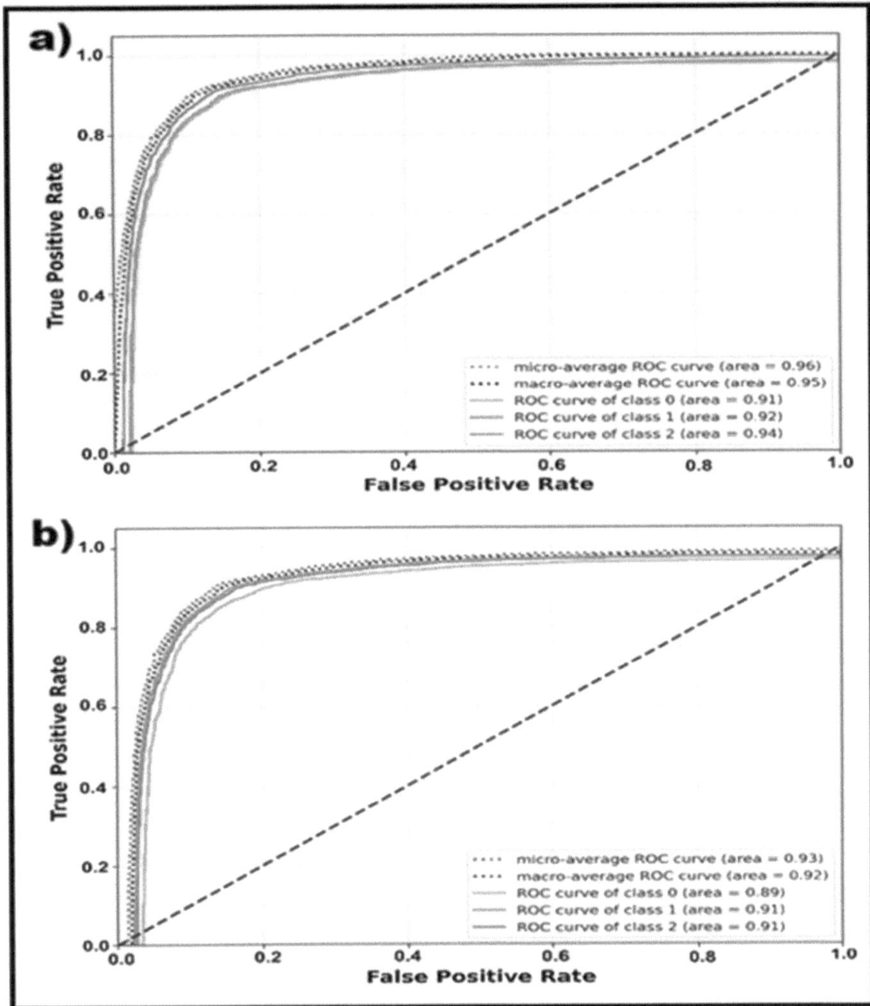

Fig. 6. Receiver Operating Characteristic (ROC) curves of the Customized CNN model for the abdomen (a) and neck (b) regions, showing excellent classification performance with micro-average AUC of 0.96 and 0.93 respectively.

4 Discussion

Current research on multiclass classification of childhood BMI categories using deep learning remains limited. This section highlights relevant studies that have explored thermal imaging for obesity assessment. Notably, Gupta et al. (2020) developed a model using recurrent neural networks (RNNs) to forecast childhood obesity risk over one-, two-, and three-year periods. Their framework integrated both static and dynamic clinical data through LSTM cells, combined

Table 3. Confusion matrix-derived metrics (TP, FP, FN, TN, Specificity, Sensitivity, Accuracy) for each model, region, and class.

Model	Region	Class	TP	FP	FN	TN	Specificity (%)	Sensitivity (%)	Accuracy (%)
Customized CNN	Abdomen	Normal	18	2	3	17	89.5	85.7	87.5
		Overweight	17	3	4	16	84.2	80.9	84.3
		Obese	20	2	2	16	88.9	90.9	89.6
	Neck	Normal	17	2	2	19	90.5	89.5	90.0
		Overweight	18	2	3	17	89.5	85.7	87.5
		Obese	19	1	3	17	94.4	86.4	89.5
DenseNet121	Abdomen	Normal	17	2	2	18	90.0	89.5	89.6
		Overweight	19	1	3	17	94.4	86.4	88.7
		Obese	20	1	1	18	94.7	95.2	95.0
	Neck	Normal	15	3	4	17	85.0	78.9	82.5
		Overweight	18	2	2	17	89.5	90.0	87.5
		Obese	19	1	2	18	94.7	90.5	92.5
Inception ResNet V2	Abdomen	Normal	16	3	4	17	85.0	80.0	82.5
		Overweight	17	3	3	17	85.0	85.0	85.0
		Obese	18	2	3	17	89.5	85.7	87.5
	Neck	Normal	15	3	4	18	85.7	78.9	82.5
		Overweight	17	2	4	17	89.5	80.9	84.3
		Obese	20	1	1	18	94.7	95.2	95.0

Table 4. Effect of optimizers and learning rates on validation and test performance for abdomen and neck regions.

Sl. No	Study Region	Optimizer	Learning Rate	Val_Acc (%)	Val_Loss	Ts_Acc (%)	Ts_Loss	AUC	Findings
1	Abdomen	Adam	0.001	92.9	0.08	93.4	0.07	0.96	Enhanced Acc.
			0.006	91.5	0.09	92.1	0.08	0.95	Accuracy Drop
		Adamax	0.001	89.8	0.11	90.3	0.10	0.91	Accuracy Drop
			0.006	87.4	0.12	88.7	0.11	0.90	Accuracy Drop
2	Neck	Adam	0.001	91.1	0.09	91.8	0.08	0.93	Enhanced Acc.
			0.006	90.4	0.09	91.1	0.09	0.94	Accuracy Drop
		Adamax	0.001	88.7	0.12	89.3	0.12	0.91	Accuracy Drop
			0.006	87.1	0.12	87.8	0.12	0.90	Accuracy Drop

with a feedforward neural network, and was trained on a large pediatric EHR dataset. Transfer learning across sub-cohorts improved generalization. The model achieved AUC scores of 0.80 (age 5), 0.93 (age 11), and 0.92 (age 18), outperforming traditional ML techniques by effectively capturing temporal patterns in health data [18]. In another relevant study, Umapathy et al. (2020) carried out a comprehensive analysis to classify obesity using thermal images from specific body regions—namely the abdomen, forearm, and shank [9]. They developed a custom deep learning model to distinguish between normal and obese individuals and compared its performance against pre-trained CNN architectures and conventional machine learning algorithms. They analyzed thermograms from 100 participants, comprising 50 obese individuals and 50 healthy controls matched

Table 5. Architectural Ablation Study of the Proposed Custom CNN

Model Variant	Description	Accuracy (%)	F1-Score	Observation
Baseline Custom CNN	Full model (Conv + BN + Dropout + Dense)	91.8	0.90	Best overall performance
Shallow CNN	One convolutional block removed	86.2	0.83	Loss of feature discrimination
No Dropout	Dropout layers removed	84.7	0.81	Overfitting, reduced generalization
Reduced Dense Units	Fewer neurons in fully connected layer	87.4	0.84	Slight drop in classification power
No Batch Normalization	BN removed from Conv blocks	83.9	0.80	Unstable training, lower robustness

by age and gender. After preprocessing and augmentation, average skin surface temperatures were compared across key body regions. Notably, the abdominal region showed a temperature difference of 4.703% between obese and normal groups. Their Custom Network-2 achieved 92% accuracy and an AUC of 0.948, outperforming other CNNs.

In the present study, thermal images from the abdomen and neck were used to classify children into normal, overweight, and obese groups using Inception-ResNetV2, DenseNet121, and a Customized CNN. The Customized CNN delivered the best results, with 90.5% accuracy, surpassing DenseNet121 (88.2%) and InceptionResNetV2 (87.1%). Despite these promising findings, certain limitations must be acknowledged. First, the number of subjects (n = 150) is relatively modest for deep learning in healthcare applications. To address this, we employed a patch-based strategy that expanded the dataset to 3,000 thermal patches, which improved the diversity of training samples and reduced the risk of overfitting. Moreover, the Custom CNN was intentionally designed as a lightweight model with fewer trainable parameters than conventional deep networks, making it more suitable for smaller medical datasets. The ablation study further confirmed the robustness of this approach, showing consistent performance across variations in hyperparameters. These outcomes demonstrate the strength of customized CNNs in recognizing subtle thermal variations linked to body composition, particularly in pediatric cases. The model's performance supports the value of lightweight, task-specific networks for multiclass classification in thermal imaging. Second, the data were collected from a single site, which may limit the generalizability of the findings to wider pediatric populations. Thermal patterns can be influenced by ethnicity, geographic location, environmental conditions, and imaging protocols. While the proposed model demonstrated robust results within the current dataset, its applicability to other cohorts remains to be validated. Future work will therefore focus on expanding data collection across multiple centers with larger and more diverse populations, which will allow us to evaluate cross-site generalizability and further enhance clinical utility.

From a clinical standpoint, the proposed thermal imaging-based approach offers several advantages. Infrared cameras are cost-effective compared to MRI or CT and require minimal operational training, making them feasible for integration into school health screening and pediatric clinics. The non-invasive and radiation-free nature of thermal imaging is particularly suited to children, ensuring comfort and safety during routine assessments. With rapid image acquisition and modest computational requirements of the customized CNN, real-time classification of obesity levels can be achieved. This provides a valuable adjunct to conventional anthropometric indices, potentially improving early detection and intervention. However, before clinical translation, larger multi-center studies and standardized acquisition protocols are needed to ensure reproducibility and generalizability.

5   Conclusion

This study utilized thermal imaging as a non-invasive approach to assess childhood obesity by capturing skin temperature data from the abdomen and neck of 150 children, equally divided into normal, overweight, and obese categories. Average temperature readings from both regions were used to train three deep learning models: a Customized CNN, a modified DenseNet121, and a modified ResNetV2. The Customized CNN achieved the highest overall classification accuracy of 92.35%, outperforming DenseNet121 (89.7%) and ResNetV2 (87.1%). Abdominal thermograms consistently produced better classification outcomes than neck images, suggesting a clearer thermal distinction in that region. These findings support the use of deep learning, particularly custom CNNs, combined with thermal imaging as an effective, contactless, and affordable method for early obesity screening in children.

While the present work focused on evaluating abdominal and neck regions independently, future research could explore integrating both anatomical sites within a unified framework. Such a feature-fusion or ensemble-based approach may capture complementary thermal information, potentially yielding higher classification accuracy and improved robustness. Expanding this line of work with larger, multi-center datasets will further enhance the generalizability and clinical applicability of the proposed method. Future work will focus on expanding the dataset to include larger and more diverse pediatric populations from multiple centers, thereby improving the robustness of the proposed model. Multimodal approaches combining thermal imaging with anthropometric or RGB data could further enhance predictive performance. Additionally, lightweight CNN variants will be explored for deployment on mobile or edge devices to facilitate school-based and community-level screenings. Longitudinal studies are also planned to evaluate the role of thermal imaging in monitoring intervention outcomes.

Acknowledgments. The authors would like to express their sincere gratitude to the Department of Biomedical Engineering, SRMIST, for providing the necessary facilities and support to carry out this research. We also thank the participating students and school authorities for their cooperation during the data collection phase.

Disclosure of Interests. The authors declare that they have no conflicts of interest relevant to the content of this article.

References

1. Mondal, P.K., Foysal, K.H., Norman, B.A., Gittner, L.S.: Predicting childhood obesity based on single and multiple well-child visit data using machine learning classifiers. Sensors **23**(2), 759 (2023)
2. Liu, Q., Li, M., Wang, W., et al.: Infrared thermography in clinical practice: a literature review. Eur. J. Med. Res. **30**, 33 (2025)
3. Singh, D., Singh, A.K., Tiwari, S.: Early thermographic screening of breast abnormality in women with dense breast by thermal, fractal, and statistical analysis. In: Singh, A.K., et al. (eds.) Artificial Intelligence over Infrared Images for Medical Applications and Medical Image Assisted Biomarker Discovery. Lecture Notes in Networks and Systems, vol. 528, pp. 25–36. Springer, Heidelberg (2022). https://doi.org/10.1007/978-3-031-19660-7_3
4. Katte, P., Kakileti, S.T., Madhu, H.J., Manjunath, G.: Automated thermal screening for COVID-19 using machine learning. In: Singh, A.K., et al. (eds.) Artificial Intelligence over Infrared Images for Medical Applications and Medical Image Assisted Biomarker Discovery. Lecture Notes in Networks and Systems, vol. 528, pp. 85–96. Springer, Heidelberg (2022). https://doi.org/10.1007/978-3-031-19660-7_7
5. Rashmi, R., Umapathy, S., Krishnan, P.: Thermal imaging method to evaluate childhood obesity based on machine learning techniques. Int. J. Imaging Syst. Technol. **31**(3), 1752–1768 (2021)
6. Leo, H., Saddami, K., Roslidar, Muharar, R., Munadi, K., Arnia, F.: Lightweight convolutional neural network (CNN) model for obesity early detection using thermal images. Dig. Health **10**, 20552076241271639 (2024)
7. Poma, Y., Melin, P., Gonzalez, C.I., Martinez, G.E.: Optimization of convolutional neural networks using the fuzzy gravitational search algorithm. J. Autom. Mobile Rob. Intell. Syst. **14**(1), 109–120 (2019)
8. An, R., Shen, J., Xiao, Y.: Applications of artificial intelligence to obesity research: scoping review of methodologies. J. Med. Internet Res. **24**(12), e40589 (2022)
9. Umapathy, S., Thanaraj, K.P., Sangamithirai, K.: Computer aided diagnosis of obesity based on thermal imaging using various convolutional neural networks. Biomed. Signal Process. Control **63**, 1–10 (2020)

10. Jiang, X., Hu, Z., Wang, S., Zhang, Y.: Deep learning for medical image-based cancer diagnosis. Cancers **15**(14), 1–72 (2023)
11. Mienye, I.D., Swart, T.G.: A comprehensive review of deep learning: architectures, recent advances, and applications. Information **15**(12), 755–800 (2024)
12. Salehi, A.W., et al.: A study of CNN and transfer learning in medical imaging: advantages, challenges, future scope. Sustainability **15**(7), 5930–5958 (2023)
13. Chai, J., Zeng, H., Li, A., Nagai, E.W.T.: Deep learning in computer vision: a critical review of emerging techniques and application scenarios. Mach. Learn. Appl. **6**(15), 1–13 (2021)
14. Zhou, T., Ye, X., Lu, H., Zheng, X., Qiu, S., Liu, Y.: Dense convolutional network and its application in medical image analysis. Biomed. Res. Int. **2022**, 2384830 (2022)
15. Chen, L., Li, S., Bai, Q., Yang, J., Jiang, S., Miao, Y.: Review of image classification algorithms based on convolutional neural networks. Remote Sens. **13**(22), 4712–4763 (2021)
16. Montaha, S., Azam, S., Rafid, A.K.M.R.H., Islam, S., Ghosh, P., Jonkman, M.: A shallow deep learning approach to classify skin cancer using down-scaling method to minimize time and space complexity. PLoS One **17**(8), e0269826 (2022)
17. Zhou, M., Li, B., Wang, J.: Optimization of hyperparameters in object detection models based on fractal loss function. Fractal Fract. **6**(12), 706–723 (2022)
18. Gupta, M., Phan, T.T., Bunnell, H.T., Beheshti, R.: Obesity prediction with EHR data: a deep learning approach with interpretable elements. ACM Trans. Comput. Healthc. **3**(3), 32–63 (2022)

A Non-invasive Diagnostic Approach to Orofacial Pain Using Infrared Thermography and Machine Learning

Nithyakalyani Krishnan(iD) and U. Snekhalatha(✉)(iD)

Department of Biomedical Engineering, College of Engineering and Technology, SRM Institute of Science and Technology, Kattankulathur, Chennai, Tamil Nadu, India
{nithyakk1,snehalau}@srmist.edu.in

Abstract. Orofacial pain is a complicated condition that frequently presents significant diagnostic challenges as a result of its complex nature and overlapping clinical presentations. In order to prevent protracted discomfort and ineffective treatment, it is essential to achieve early and accurate detection. This study suggests a non-invasive diagnostic method that combines infrared thermal imaging with machine learning techniques to differentiate between normal and abnormal cases of orofacial pain. Thermal facial images were acquired under standardized conditions, pre-processed, and subjected to a comprehensive feature extraction process that included statistical, textural, spatial, and raw pixel properties. These were standardized, dimensionality-reduced using Principal Component Analysis (PCA), and classified using Decision Tree, XGBoost, Random Forest, and Naive Bayes algorithms. In order to promote a comprehensive assessment of the accuracy, precision, recall, F1 score, and Area under receiver operating characteristics (AUC) metrics, a stratified 5-fold cross-validation was implemented. The Decision Tree exhibited the highest accuracy among the other models, demonstrating its capacity to accurately identify all true positive cases without missed diagnoses while maintaining high precision. The PCA projection demonstrated a distinct distinction between normal and aberrant cases, with classifiers that are capable of navigating non-linear boundaries. The results indicate that the integration of interpretable machine learning models with thermal imaging can yield a clinically viable, objective, and dependable tool for the detection of orofacial pain. This tool has the potential to be integrated into clinical decision support systems.

Keywords: Orofacial pain · infrared thermography · machine learning · principal component analysis · disease detection

1 Introduction

Orofacial pain refers to discomfort originating from the oral and facial areas. This may result from nervous system failure, local anatomical abnormalities, or pain

© The Author(s), under exclusive license to Springer Nature Switzerland AG 2026
S. T. Kakileti et al. (Eds.): AIIIMA 2025, LNCS 16308, pp. 113–127, 2026.
https://doi.org/10.1007/978-3-032-10990-3_8

referred from remote locations [1]. Orofacial pain can be categorized as acute or chronic. Generally acute pain correlates with discernible tissue damage and usually subsides with suitable intervention. Orofacial pain lasting more than three months is classified as chronic [2]. This encompasses various conditions, with the most common issues being dental disorders like toothaches and periodontal infections, temporomandibular joint disorders (TMJ), and neuropathic pain conditions such as trigeminal neuralgia. Other causes include musculoskeletal issues, nasal difficulties, and cephalalgia or migraines [3].

The accurate diagnosis of orofacial pain is crucial for effective management, due to its complex etiology and overlapping clinical features. Misdiagnosis or delayed identification often leads to ineffective treatments, prolonged patient distress, and psychological damage. A systematic diagnostic process includes a comprehensive medical and dental history, clinical evaluation, imaging studies, neurophysiological tests, and psychological evaluations. The clinical assessment begins with an extensive medical history that highlights the initial onset, duration, quality, and location of pain, along with triggering and relieving factors, radiation patterns, and relevant history of trauma, dental procedures, or systemic conditions. The physical examination includes inspection and palpation of the masticatory muscles and temporomandibular joints, assessment of mandibular range of motion, intraoral examination for dental or mucosal pathology, and neurological evaluation of the cranial nerves, particularly the trigeminal nerve. The diagnosis accuracy of musculoskeletal discomfort can be improved with the application of standardized tools, such as the diagnosis Criteria for Temporomandibular Disorders (DC/TMD).

Imaging studies are crucial for the differential diagnosis of orofacial discomfort. Magnetic resonance imaging (MRI) is considered the gold standard for analyzing soft tissue components, such as the TMJ articular disc and associated muscles, whereas panoramic radiographs provide a first evaluation of dental and skeletal structures. The assessment of degenerative changes in the TMJ is significantly enhanced by cone-beam computed tomography (CBCT), which produces detailed images of bony structures. Ultrasound imaging serves as a non-invasive and cost-effective method for evaluating superficial muscles and joint cavities. In recent years, infrared thermal imaging has emerged as a significant tool for assessing orofacial pain. This non-invasive, radiation-free technique detects surface temperature variations that may indicate underlying inflammation, neuropathic conditions, or vascular anomalies. Its capacity to accurately assess facial thermal asymmetry and track dynamic alterations renders it especially important for clinical research and the advancement of AI-driven diagnostic tools.

The suggested approach employs machine learning algorithms, including Decision Tree, XGBoost, Random Forest, and Naive Bayes, to enhance the identification and categorization of orofacial pain through thermal imaging. This entails preprocessing thermal face photos, collecting pertinent features that encapsulate statistical, textural, spatial, and raw thermal patterns, and subsequently employing supervised learning algorithms to differentiate between normal and pathological thermal signatures. This amalgamation of thermal imaging and machine learning seeks to deliver a non-invasive, automated, and objective instrument for aiding in the identification of orofacial pain disorders.

2   Related Works

Researchers have demonstrated improved sensitivity and specificity in disease identification by the integration of thermal imaging with machine learning models, including Support Vector Machines, Random Forest, Gradient Boosting, and deep learning methodologies such as Convolutional Neural Networks (CNNs). Machine learning-enhanced thermography has been efficiently employed in breast cancer screening, early detection of diabetic foot ulcers, and assessment of orofacial pain. Machine learning and thermal imaging together provide an objective, automated, and reproducible framework that reduces human subjectivity, enables early diagnosis, and supports large-scale population screening.

Recent research have proved the efficacy of thermal imaging combined with artificial intelligence for breast cancer screening. Real-time thermography was utilized alongside Inception-based neural models to attain remarkable classification accuracy. An altered Inception Mv4 attained a 99.7% accuracy rate. The research showed that in situ chilling improved image quality, and the efficacy of detection was significantly improved by the presence of minimal fluctuations in tumor surface temperature. These findings highlight the therapeutic potential of integrating thermal imaging with advanced deep learning techniques for early cancer detection [5]. A proof-of-concept study applied machine learning to thermal and RGB images for assessing surgical wounds. The low infection rate constrained the identification of surgical site infections; however, the models attained 89 to 92% accuracy in wound segmentation, underscoring the promise of thermal imaging and computer vision for postoperative monitoring [6].

In recent years, there has been substantial progress in the automated diagnosis of orofacial pain using facial images. Umapathy et al. achieved exceptional classification accuracies of up to 99% with Support Vector Machine (SVM) and Random Forest, and 97% with VGG-16 in the field of thermography by implementing a dual approach that combined classical machine learning and deep learning on facial thermograms [7]. This highlights the potential of both hand-crafted and automated feature strategies. The value of combining semantic and radiomic feature extraction from infrared images under standardized conditions was demonstrated in similar clinical studies, such as those on TMDs, thereby reinforcing the feasibility of infrared quantitative thermography in orofacial pain contexts [8].

In addition to thermal imaging, a comprehensive review of the automated pain assessment of visible-spectrum facial images reveals a substantial body of research that employs ensemble methods, convolutional neural networks, and CNN-RNN hybrids. For instance, Semwal and colleagues obtained an accuracy of up to 96% on benchmark datasets such as UNBC-McMaster by combining VGG-based models with texture features [9]. Sikdar et al. highlighted thermography as a non-contact, radiation-free imaging modality based on infrared emission from the skin. Since skin temperature reflects metabolic, vascular, and neural changes, the technique offers potential in detecting pathological alterations. In dentistry, the authors reviewed its history and underscored its diagnostic applications, particularly for inflammatory, vascular, temporomandibular joint, and neoplastic conditions. They recommended thermography as a valuable adjunct for early diagnosis and monitoring in oral medicine [10].

Real-time clinical monitoring has been facilitated by the implementation of deep learning architectures, such as ResNet-based systems, which autonomously classify pain expressions [11]. The inclusion of both deep learning and hand-crafted feature approaches in automated pain recognition is essential, as each approach provides distinct advantages: Deep learning is particularly effective when dealing with large, diverse datasets, despite the problems of fairness and explainability, whereas handcrafted methods offer interpretability and reduced data requirements [12]. AI-based pain assessment systems, such as those that employ ResNet-18 to classify postoperative pain, demonstrated superior sensitivity and specificity in comparison to human medical personnel during the clinical evaluation. This suggests a substantial potential for real-world application [13]. The importance of standardized and validated imaging pipelines, the efficacy of integrating feature engineering and machine learning, and the necessity of addressing issues of interpretability, dataset diversity, and impartiality in pain detection systems are underscored by these studies.

3   Materials and Methods

In this section, a comprehensive examination of the study overview illustrated in Fig. 1 will be presented. Documentation of the experimental procedures, which include data acquisition and subjects, is initiated. As a result, the criteria for feature selection and data preprocessing are established. A comprehensive description of the machine learning methodologies that are implemented in the functions of data analysis and classification is offered.

3.1   Participants and Ethical Approval

The purpose of this study is to evaluate the clinical feasibility of thermal imaging and machine learning in the detection of orofacial pain. Normal subjects and patients with orofacial pain were recruited from SRM Kattankulathur Dental College & Hospital in Tamil Nadu, India. Ultimately, it was determined that 40 subjects aged 25 to 50 who experienced orofacial pain and 40 subjects who were normal were recruited for this investigation. The voluntary informed consent of all participants in this study was obtained. This investigation was granted ethical approval by the Institutional Ethics Committee (IEC) of SRM Medical College Hospital and Research Centre, Kattankulathur, Tamil Nadu, India (Ethics Clearance Number: 1343/IEC/2018).

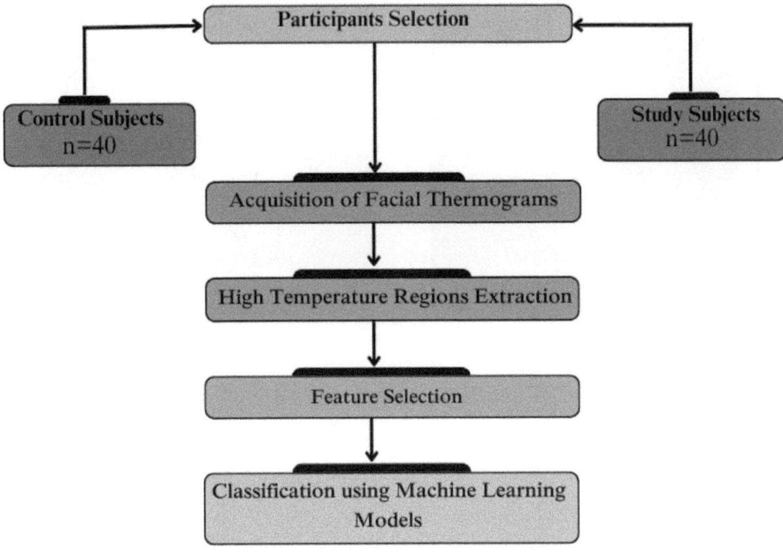

Fig. 1. Overview of the methodology adopted in this research study.

3.2   Infrared Imaging Protocol

Facial thermograms were obtained using a FLIR A305sc thermal camera connected to a computer operating FLIR Tools software, which quantifies infrared radiation and produces thermal images where each pixel denotes a precise temperature value. The camera was affixed to a tripod at a distance of 1 m from the participant and permitted to calibrate for 5 to 10 min before imaging. To facilitate adequate acclimatization, participants remained in a controlled environment space for 10–15 min, with the temperature set at 20° C and airflow regulated to avoid disruptions. Subsequent to this interval, participants were instructed to adopt a comfortable posture while gazing in three orientations: straight ahead, to the right, and to the left.

The study excluded participants who were adorned with ornaments or jewellery that obstructed the area of interest and could not be removed, or who had facial hair. Thermal images were acquired for the anterior, left, and right profiles of each participant. The participant was positioned behind a black background to get rid of reflections and thermal noise, improve contrast, standardize imaging conditions, reduce heat interference from outside sources, and focus the infrared camera on the subject [14]. Figure 2 represents the sample thermal images of the control and study subjects.

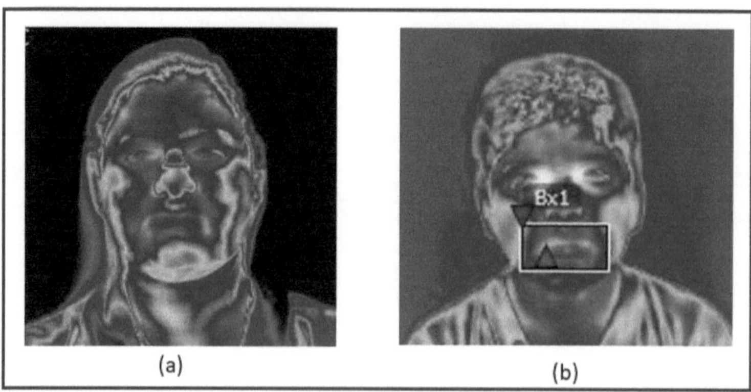

Fig. 2. Facial Thermograms (a) Normal subject, (b) Subject with orofacial pain, emphasizing the region of interest (ROI) that indicates an aberrant temperature distribution on the face.

3.3   Proposed Methodology

Data Loading and Preparation. This study utilized images from two specified directories labelled as "normal" and "abnormal" representing healthy and pathological cases, respectively. Image preprocessing was conducted leveraging the OpenCV package to ensure uniformity and relevance for machine learning analysis. Each image was first loaded using the imread() function in OpenCV. To diminish computational complexity and concentrate on intensity patterns pertinent to thermal imaging, the images were transformed to grayscale utilizing appropriate parameters. The input dimensions remain consistent throughout the dataset by enlarging all of the images to a uniform dimension of 128×128 pixels using the resize() function.

In neural network models that require fixed input dimensions, the resizing procedure is indispensable for effective training and inference. Furthermore, fundamental validation procedures have been implemented to verify the quality of the data. This entailed the identification and elimination of uninformative images, specifically those that were completely black, white, or exhibited minimal pixel intensity variation. These images lack substantial substance and may adversely affect model performance. Additionally, the dataset has been cleared of files that were either inaccessible or defective. This preprocessing workflow verified that the input images were homogenous, unaltered and suitable for future analysis.

Feature Extraction. Various types of features were extracted from each preprocessed thermal image during the feature extraction phase to accurately represent the underlying patterns. The thermal distribution of the image was initially assessed by calculating the mean, standard deviation, median, maximum, and minimum pixel intensities. The Local Binary Pattern (LBP) technique was then employed to derive texture attributes, which is capable of highlighting complex textural variations. Local texture differences that may indicate anomalous thermal signatures were captured by generating a histogram for each image that included 20 bins of LBP values.

Additionally, spatial characteristics were incorporated by dividing each image into four quadrants and determining the mean pixel intensity for each quadrant. This spatial segmentation is instrumental in identifying aberrant conditions by capturing regional temperature variations. The unprocessed pixel values of each $128{\times}128$ image were ultimately condensed into a single vector, ensuring that the image retains detailed intensity information. A singular feature matrix was constructed by combining the extracted statistical, textural, spatial, and raw pixel values. Consequently, this matrix was utilized as input for machine learning models during the analysis phase. This method is comprehensive and verifies the accurate representation of both global and localized thermal properties.

Standardization and Dimensionality Reduction. In order to optimize the extracted features for machine learning, a dimensionality reduction and standardization procedure was implemented. Initially, the StandardScaler function from the scikit-learn library was implemented to scale features. This step guarantees that all features contribute equally to the learning process and precludes any dominance by features with larger numerical ranges by standardizing the features by removing the mean and scaling them to unit variance. The dimensionality of the feature space was reduced through the application of Principal Component Analysis (PCA) after standardization, thereby retaining the most significant information. PCA identifies the principal components (the directions) that account for the most variance in the data. Subsequently, the feature vectors are projected onto these new axes. The number of principal components was determined to guarantee that 95% of the total variance in the original data was preserved. In addition to reducing computational complexity, this also facilitates the elimination of redundant information and noise, thereby enhancing the generalization and efficacy of the model. Subsequently, the lower-dimensional feature set was employed to train and assess the machine learning models.

Model Selection. Thermal images were classified using various types of machine learning models to evaluate their performance in a range of algorithmic frameworks. The interpretability and capacity to capture non-linear feature interactions of Decision Tree were selected due to their utility in clinical decision-making. The data is divided according to an impurity criterion, such as the Gini index or Entropy, and is represented in Eq. (1).

$$G = 1 - \sum p_i^2 \tag{1}$$

where G is Gini impurity and p_i = Probability of class i in a node.

Random Forest was incorporated as an ensemble method that enhances generalization and reduces overfitting by bootstrapping multiple decision trees systematically. It is an ensemble of Decision Trees using bagging and feature randomness. The final prediction of random forest is expressed in Eq. (2) where $f_{RF}(x)$ is the final prediction of the random forest, $f_b(x)$ is the prediction of the b-th decision tree and B is the total number of trees.

$$f_{RF}(x) = \frac{1}{B} \sum_{b=1}^{B} f_b(x) \tag{2}$$

XGBoost was chosen due to its efficacy, regularization mechanisms, and demonstrated success in managing structured tabular datasets. This algorithm sequentially constructs trees by minimizing an objective function with regularization, and the expression is illustrated in Eq. (3).

$$\text{Obj}(\theta) = \sum l(y_i, \hat{y}_i) + \sum \Omega(f_k) \tag{3}$$

where $\text{Obj}(\theta)$ is the objective function, $l(y_i, \hat{y}_i)$ is the loss function between true value y_i and predicted value $\hat{y}_i$ and $\Omega(f_k)$ is the regularization term penalizing tree complexity.

Naive Bayes was considered as a probabilistic baseline due to its simplicity, computational efficiency, and robustness in small datasets, serving as a benchmark for comparison. Equation (4) represents the posterior probability, which is derived from Bayes' theorem and assumes feature independence.

$$P(C|X) = \frac{P(X|C)P(C)}{P(X)} \tag{4}$$

where $P(C|X)$ is the posterior probability of class C given features X, $P(X|C)$ is the likelihood of observing X given class C, $P(C)$ is the prior probability of class C and $P(X)$ is the evidence (overall probability of features X).

Initially, the Gaussian Naive Bayes (GNB) classifier was implemented. This probabilistic model is based on Bayes' theorem and assumes that the features follow a Gaussian distribution. The method's optimal foundation is that of speed and efficacy, making it particularly well-suited for high-dimensional data. The data was partitioned according to feature levels using a decision tree classifier, which creates a tree-like model that is comprehensible. It is especially proficient at effectively capturing non-linear correlations and interactions between features.

Overfitting was also reduced and performance was improved by using an ensemble technique called Random Forest. This model utilizes various data subsets to build decision trees, which are then averaged to improve the model's predictive capacity and dependability. They then used XGBoost, a powerful gradient boosting method. Because of its lightning-fast speed and performance on structured data jobs, XGBoost employs complex regularization methods to ensure that the models it generates are as correct as possible and creates trees in a certain order. The integration of different models allowed for a thorough evaluation of each one in terms of classification accuracy, model resilience, and computing efficiency.

4   Results

4.1   Model Evaluation

This study employed a stratified 5-fold cross-validation strategy to objectively evaluate the accuracy of the model and ensure the evaluation process's reliability and integrity. In an effort to mitigate potential class imbalances, the dataset was divided into five equal subsets, which were termed "folds." The original class distribution was maintained within each fold. Four folds were employed for each iteration during the training phase, while the remaining fold was for testing. Every single occurrence in the dataset was utilized as a test sample exactly once inside each of the five iterations of this method. In order to achieve reliable and objective estimations of the prediction capacity of each model, the performance ratings that were collected from all folds were then averaged.

Multiple evaluation measures were utilized in order to completely understand the effectiveness of the model. While the evaluation metric, area under receiver operating characteristics (AUC) estimated the ability of the model to differentiate between normal and abnormal cases across a variety of decision thresholds, accuracy measured the overall proportion of right predictions across the entire process. In contrast to recall, which tested the capacity of the model to accurately identify all positive cases, precision examined the proportion of samples that were categorized as positive (abnormal) even if they were truly positive. A balanced measurement was supplied by the F1-score, which is the harmonic mean of precision and recall. This has proven to be particularly relevant in situations when false positives and false negatives carry equivalent clinical repercussions. Through the utilization of stratified cross-validation in conjunction with these several performance indicators, the evaluation framework was able to guarantee that the diagnostic capability of every machine learning model was evaluated in a manner that was reliable, objective, and clinically relevant.

4.2   Comparative Evaluation

Key performance metrics, including accuracy, F1 score, precision, recall, and AUC, were employed to assess the performance of four machine learning classifiers such as Decision Tree, XGBoost, Random Forest, and Naive Bayes. As shown in Table 1, the decision tree classifier outperformed XGBoost, Random Forest, and Naive Bayes, achieving the highest accuracy among all the evaluated models. It also achieved perfect recall, indicating that all true cases of orofacial pain were correctly identified without any missed diagnoses. Its high precision reflects a low rate of false positives, which is crucial for reducing unnecessary clinical follow-ups and avoiding patient anxiety caused by incorrect alerts. The high F1 score demonstrates an excellent balance between sensitivity and precision. While Random Forest and XGBoost achieved near-perfect receiver operating characteristics (ROC) AUC values, indicating strong class separability, their slightly lower accuracy suggests a marginally higher rate of misclassification. Naive Bayes, although showing acceptable recall and ROC AUC, had noticeably lower accuracy and precision, suggesting a higher tendency for false positives. The ROC plot of all the utilized machine learning models are depicted in Fig. 3.

The interpretability of the Decision Tree provides a particular advantage for clinical adoption, as its decision criteria are transparent and easily confirmed by healthcare professionals. This is in addition to the fact that the Decision Tree performs well numerically. Integrating the Decision Tree model into clinical decision support systems for orofacial pain detection is the best option due to its high diagnostic performance, explainability, and lack of missing cases.

The two-dimensional visual representation of the final extracted features used by the machine learning classifiers to distinguish between normal and aberrant cases is provided by the PCA projection depicted in Fig. 4. The normal samples (blue) and aberrant samples (orange) exhibit a moderate degree of separation, with normal cases clustering more tightly towards the left side of the plot, while abnormal cases are more dispersed towards the center and right. This is evident in the graph. This separation implies that the features selected and reduced through PCA retain discriminative power, which explains the robust performance of the classifiers, particularly those that effectively manage non-linear decision boundaries (Decision Tree, Random Forest, XGBoost).

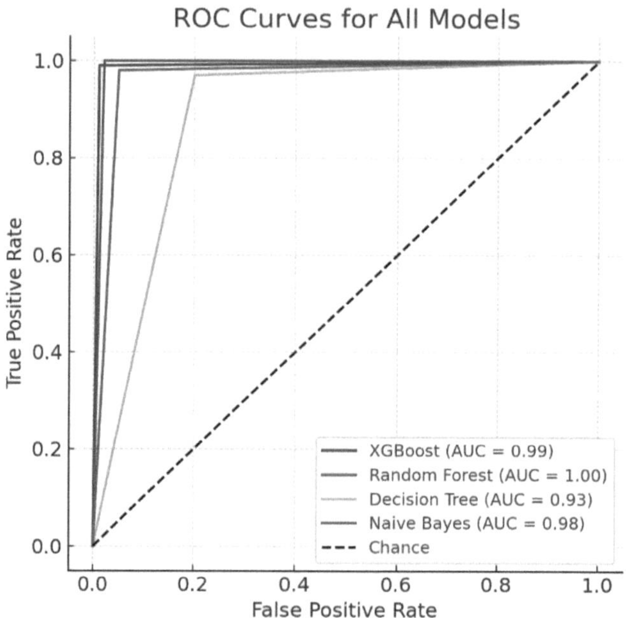

Fig. 3. Receiver Operating Characteristic (ROC) curves showing model performance for orofacial pain classification.

Table 1. Comparative analysis of various machine learning classifiers for the detection of orofacial pain.

Model	Accuracy	ROC AUC	F1	Precision	Recall
Decision Tree	0.957	0.950	0.968	0.945	1.000
XGBoost	0.928	0.995	0.946	0.903	1.000
Random Forest	0.927	1.000	0.943	0.890	1.000
Naive Bayes	0.826	0.964	0.878	0.804	0.977

Classifiers are able to achieve high recall for this category as a result of the tighter grouping of normal samples, which indicates lower intra-class variability. Conversely, the increased heterogeneity within the abnormal class is suggested by the broader distribution of abnormal samples. This could potentially present a modest challenge for certain models in terms of precision, as Naive Bayes exhibits lower precision than tree-based models.

In conclusion, the PCA-separated clusters align well with distinct, interpretable decision boundaries, which is why Decision Tree performed the best when compared to the other utilized ML models. Random Forest and XGBoost also effectively managed the non-linear separations, as evidenced by their high ROC AUC values. The overlap regions were more challenging for Naive Bayes, resulting in a decrease in accuracy and precision. This is likely due to the assumption of feature independence.

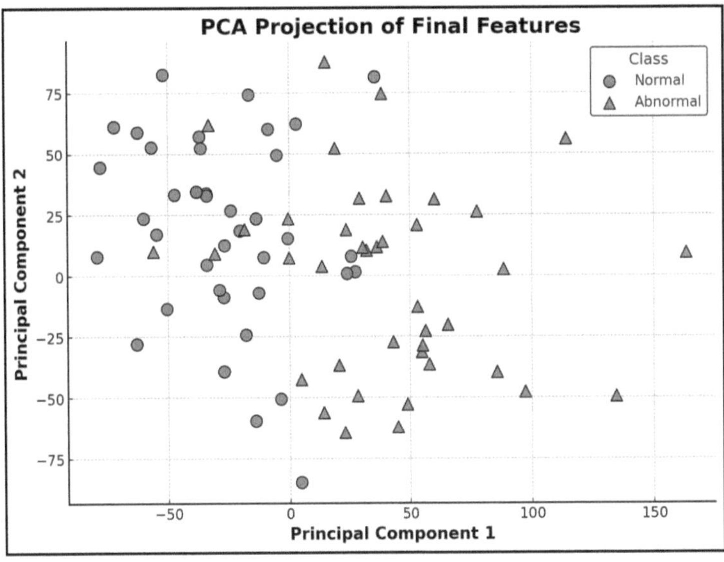

Fig. 4. PCA Projection of Final Features for Classifying Normal and Abnormal Cases.

5   Discussion

The present study demonstrates the potential of infrared thermography (IRT) combined with machine learning (ML) as a non-invasive approach for the diagnosis of orofacial pain. Our findings suggest that thermal imaging can capture physiologically relevant temperature variations associated with orofacial pain, and when analyzed through ML models, these variations can be leveraged for diagnostic decision-making. Table 2 represents the detailed summary of this research study.

Table 2. Summary of the study highlighting dataset characteristics, applied methodologies, key results, limitations, and future directions for developing a non-invasive diagnostic framework for orofacial pain using infrared thermography and machine learning.

Aspect	Details
Objective	To develop a non-invasive diagnostic framework for orofacial pain using infrared thermography (IRT) and machine learning
Dataset	80 participants (n = 80) Normal and abnormal cases of orofacial pain
Features Extracted	Thermal asymmetry, mean/variance in ROIs, statistical and texture features
Dimensionality Reduction	Principal Component Analysis (PCA)
Models Used	Decision Tree, Random Forest, XGBoost, Naïve Bayes
Best Performance	Random Forest (AUC = 1.00), XGBoost (AUC = 0.99), Naïve Bayes (AUC = 0.98)
Limitations	Small sample size (n = 80), no external validation, risk of overfitting
Clinical Relevance	Demonstrates potential of IRT + ML as a non-invasive diagnostic tool for orofacial pain
Future Work	Expand dataset, perform external validation, improve generalizability, test real-time clinical applicability

Among the utilized models, the Decision Tree achieved the highest classification performance. This finding highlights the ability of simple, interpretable models to provide clinically useful outputs in diagnostic settings. The strength of this approach lies in its non-invasive nature, absence of radiation exposure, and relatively low cost compared to other imaging modalities. In addition, the integration of ML enhances diagnostic accuracy beyond what can be achieved through visual thermal pattern analysis alone.

While the PCA-based dimensionality reduction employed in this study was effective in capturing the most discriminative features and improving model performance, it reduces interpretability, as the transformed components are linear combinations of the original features. In clinical settings, such "black-box" transformations can be difficult for healthcare professionals to interpret and trust. Therefore, although PCA enhances computational efficiency and classifier performance, there is a trade-off with transparency. Future work could explore feature selection methods that maintain interpretability while preserving discriminative power, thereby improving adoption in clinical decision support systems.

Despite these promising results, the study has certain limitations. A formal sample size calculation was performed using the formula represented in Eq. (5).

$$n = \frac{Z^2 p(1-p)}{d^2} \tag{5}$$

where n is the required sample size, $p = 0.05$ (expected prevalence), $d = 0.10$ (desired precision) and Z is the desired confidence level in a standard normal distribution. For a 95% confidence level, $Z = 1.96$. This calculation resulted in $n = 80$. However, the dataset remains relatively modest, which may limit the generalizability of the findings. External validation could not be conducted due to the absence of publicly available datasets for this condition. Although the Decision Tree model showed strong performance on the current dataset, its predictive capability should be further evaluated on independent data when available.

Future work should aim to expand the dataset to include larger, more diverse, and multi-center cohorts. Performing external validation once suitable data become accessible will be crucial for establishing robustness, reliability, and clinical applicability. Additionally, investigating more advanced machine learning approaches, such as ensemble methods or deep learning, may enhance performance while maintaining interpretability for clinical decision-making. Although our preliminary study demonstrates high recall and precision for the Decision Tree model, some misclassifications are still possible. We emphasize that ensuring patient safety and appropriate safeguards is critical, and future work will focus on strategies to reduce misclassification, including threshold optimization, uncertainty estimation, and validation on independent datasets.

A potential limitation of the current study is that orofacial pain can arise from multiple underlying causes, including dental issues, TMJ disorders, and neural or musculoskeletal conditions. Each of these causes may exhibit distinct thermal signatures or physiological patterns, which our current machine learning model does not differentiate. As a result, while the model performs well in distinguishing between normal and abnormal cases, its diagnostic specificity for different subtypes of orofacial pain may be limited. Future studies will aim to expand the dataset with annotated subcategories of orofacial pain, enabling the development of models capable of distinguishing between different etiologies, thereby improving clinical applicability and specificity.

6   Conclusion

The clinical potential and feasibility of integrating infrared thermal imaging with machine learning for the non-invasive diagnosis of orofacial pain are represented in this study. The Decision Tree classifier was the most effective model, obtaining the highest accuracy, perfect recall, and balanced precision. This ensured that no true cases were missed and that false positives were greatly reduced. Both early detection and diagnostic reliability are indispensable in clinical practice, which is why these attributes are so important. The PCA-based visualization verified that the extracted features maintained a robust discriminative capacity, which was especially advantageous for models that capture non-linear relationships. Random Forest and XGBoost also demonstrated robust ROC AUC values; however, their slightly lower accuracy underscores the Decision Tree's superior consistency in classification. Although Naive Bayes was efficient, it was less effective in managing overlapping feature regions. Ultimately, this work helps healthcare professionals manage orofacial pain in a timely and effective manner by establishing a transparent, explainable, and accurate diagnostic framework that can be easily incorporated into clinical workflows.

Acknowledgements. The Authors would like to express their sincere gratitude to SRM Kattankulathur Dental College & Hospital for the facility provided to acquire data in hospital.

References

1. Ghurye, S., McMillan, R.: Orofacial pain – an update on diagnosis and management. Br. Dent. J.(2017)
2. Zakrzewska, J. M.: Facial pain. In: Evidence-based Chronic Pain Management, pp. 134–150 (2010)
3. Labanca, M., Gianó, M., Franco, C., Rezzani, R.: Orofacial pain and dentistry management. Diagnostics **13**(17), 2854 (2023)
4. Fricova, J., Janatova, M., Anders, M., Albrecht, J., Rokyta, R.: Thermovision: a new diagnostic method for orofacial pain? J. Pain Res. **11**, 3195–3203 (2018)
5. Al Husaini, M.A.S., Habaebi, M.H., Islam, M.R.: Real-time thermography for breast cancer detection with deep learning. Disc. Artif. Intell. **4**, 57 (2024)
6. Prey, B.J., Colburn, Z.T., Williams, J.M., et al.: The use of mobile thermal imaging and machine learning technology for the detection of early surgical site infections. Am. J. Surg. **231**, 60–64 (2024)
7. Umapathy, S., Krishnan, P.T.: Automated detection of orofacial pain from thermograms using machine learning and deep learning approaches. Expert Syst. **38**(7), e12747 (2021)
8. Diniz de Lima, E., Souza Paulino, J.A., Lira de Farias Freitas, A.P., et al.: Artificial intelligence and infrared thermography as auxiliary tools in the diagnosis of temporomandibular disorder. Dentomaxillofac. Radiol. **51**(2), 20210318 (2022)
9. Ben Aoun, N.: A review of automatic pain assessment from facial information using machine learning. Technologies **12**(6), 92 (2024)
10. Sikdar, S.D., Khandelwal, A., Ghom, S., Diwan, R., Debta, F.M.: Thermography: a new diagnostic tool in dentistry. J. Indian Acad. Oral. Med. Radiol. **22**(4), 206–210 (2010)
11. Shuang, Y., Liangbo, G., Huiwen, Z., et al.: Classification of pain expression images in elderly with hip fractures based on improved ResNet50 network. Front. Med. **11**, 1421800 (2024)

12. Chiavaccini, L., Gupta, A., Chiavaccini, G.: From facial expressions to algorithms: a narrative review of animal pain recognition technologies. Front. Vet. Sci. **11**, 1436795 (2024)
13. Fontaine, D., Vielzeuf, V., Genestier, P., et al.: Artificial intelligence to evaluate postoperative pain based on facial expression recognition. Eur. J. Pain **26**(6), 1282–1291 (2022)
14. Liu, Q., et al.: Infrared thermography in clinical practice: a literature review. Eur. J. Med. Res. **30**, 33 (2025)

Thermal Asymmetry in Football Players Following Ankle Injury: Findings Related to Training Load

Ahmet Bayrak[1]([envelope]) [iD], Mahmut Çevik[2] [iD], and Murat Ceylan[3] [iD]

[1] Vocational School of Health Science, Selcuk University, Konya, Turkey
ahmetbayraklaw@gmail.com
[2] AIVISIONTECH Electronic Software Inc., Konya, Turkey
[3] Faculty of Engineering and Natural Sciences, The Department of Electrical and Electronics Engineering, Konya Technical University, Konya, Turkey

Abstract. *The Objective:* This study aimed to evaluate, at two time points (the 1st and 14th training days), the effects of training provocation on thermal asymmetry in football players with and without a lateral ankle sprain (LAS) injury history.

Methods: Twenty-seven football players from the U-19 squad of a Turkish Süper Lig club were included. Athletes were divided into two groups by injury history: with LAS injury history (n = 10) and without LAS injury history (n = 17). On the 1st and 14th training days, pre-training and post-training infrared thermographic images were obtained. For the ankle, patellar tendon, calf medialis, calf lateralis, and tibialis anterior regions, the side-to-side temperature difference (ΔT) and the post-training change (ΔPost–Pre) were calculated.

Results: On the first day, athletes with LAS injury history showed a marked increase in ΔT at the ankle (+0.19 °C) and patellar tendon (+0.22 °C), whereas a decrease was observed in the control group. On the fourteenth day, ΔT values were elevated from pre-training in the injury-history group and expanded toward the calf muscles (calf medialis +0.07 °C; calf lateralis +0.06 °C). Tibialis anterior exhibited a decrease in both groups.

Conclusion: In football players with LAS injury history, thermal asymmetry that emerges acutely at the joint–tendon level with training load extends to the muscle level over two weeks. This pattern indicates lasting alterations in neuromuscular control and load distribution. AI-assisted thermography can sensitively reveal such asymmetries and may serve as a valuable tool for individualized load management and injury prevention strategies during return-to-play.

Keywords: Thermography · Ankle Sprain · Football Injury

1 Introduction

The ankle joint is among the joints most susceptible to injury in athletes [1]. In the literature, ankle sprains are reported to account for approximately 40% of all sports injuries [2].

© The Author(s), under exclusive license to Springer Nature Switzerland AG 2026
S. T. Kakileti et al. (Eds.): AIIIMA 2025, LNCS 16308, pp. 128–142, 2026.
https://doi.org/10.1007/978-3-032-10990-3_9

Ankle sprains most commonly occur as inversion-type injuries and are characterized by overstretching or tearing of the ligaments on the lateral aspect of the ankle [3–5]. Approximately 75% of all sprains develop via an inversion mechanism, and about 73% of these involve the anterior talofibular ligament [3–7]. Such injuries are observed more frequently in sports requiring intensive running and jumping, such as football, basketball, and volleyball [4, 6–10]. The high incidence of acute ankle sprains is associated with an increased risk of recurrence following the initial injury [6]. Ankle sprains affect approximately 1.7 billion people worldwide [11] and are reported in two million individuals annually in the United States alone [12]. Regarding daily incidence, it is stated that one out of every 10,000 people sustains a lateral ankle sprain, making it the most common type of sport-related injury [13].

Given their importance in sports injuries and their propensity for ongoing complications after the initial event, the prevention of ankle sprains is of paramount importance [14, 15]. A previous ankle sprain constitutes the most commonly reported injury risk factor [16–19]. The rate of progression to chronic ankle instability is high; chronic ankle instability characterized by persistent, frequently recurring symptoms, recurrent sprains, and subjective ankle instability is observed in 10–40% of cases [20–23].

Lateral ankle ligament injuries are among the most frequent injuries in professional football. The average time loss of 15 days is shorter than the duration required for physiological wound healing. The high recurrence rate of 17% raises the question of whether the rehabilitation period is sufficient to progressively restore sport-specific function to preinjury levels [24]. Therefore, examining the impact of lateral ankle sprains as an injury risk factor is important. Evaluating whether a previous ankle sprain injury history predisposes to recurrence or contributes to injuries in other regions is of great significance for the development of injury prevention programs. However, there is a paucity of studies on this topic in the literature. In the present study, we aimed to elucidate how injury history in football players with lateral ankle sprain affects thermographic markers.

2 Method

Data were collected from 27 football players aged 19 from the U-19 squad of a Turkish Super League club. On two dates—the 1st training day of the camp period and the 14th training day of the preseason camp—infrared thermal images were obtained pre-training and post-training for each athlete. During the training camp, players followed a structured program consisting of daily morning and afternoon sessions. Morning sessions primarily emphasized aerobic conditioning, mobility, and preventive strengthening, while afternoon sessions focused on tactical drills and high-intensity small-sided games. Training load was periodized to include two high-intensity days followed by one low-intensity recovery day.

Two groups were defined by injury history:

Injury history (IH+): n = 10 athletes who had previously sustained an ankle injury resulting in ≥3 days away from play.

Injury history (IH−): n = 17 athletes with no ankle injury.

Thermal data were obtained using the **aivisiontech–ai4sports** artificial intelligence–based injury risk analysis software. The following regions of interest (ROIs) were analyzed: ankle, patellar tendon, Calf medialis, Calf lateralis, and Tibialis anterior. The thermal imaging device was FLIR T540-EST, featuring 464×348 IR resolution, a 7.5–14 μm spectral range, a 24° lens, thermal sensitivity NETD < 40 mK at 30 °C, and measurement accuracy of ± 2 °C or $\pm 2\%$ of reading.

For each ROI, day, and measurement time (pre-training/post-training), **the right–left temperature difference** (asymmetry) was computed, and all differences were treated as **absolute values.**

$$\Delta T_{r,g,z} = \left| T^{right}_{r,g,z} - T^{left}_{r,g,z} \right| \tag{1}$$

Here, r = region (ROI), g = day (1 or 14), z = time (pre-training / post-training). Training-provocation change (ΔPost–Pre):

$$\Delta Post - Pre_{r,g} = \Delta T_{r,g,after} - \Delta T_{r,g,before} \tag{2}$$

A positive value indicates an increase in asymmetry after training, whereas a negative value indicates a decrease. Analyses were reported considering group (injury history $(+/-)$) $\times$ day (1/14) $\times$ time (pre/post).

For each distribution, the mean ($\bar{x}$), standard deviation (σ) and coefficient of variation (CV, %) were calculated.

$$\sigma = \sqrt{\frac{1}{n} \sum_{i=1}^{n} (x_i - \bar{x})^2} \tag{3}$$

$$CV = \begin{cases} 100 x \frac{\sigma}{\bar{x}}, & \bar{x} \neq 0 \\ 0, & \bar{x} = 0 \end{cases} \tag{4}$$

3 Results

The distributions of right–left temperature differences (ΔT) measured on the 1st and 14th training days of a football team's summer camp period exhibited different patterns across regions. On the 1st day, a post-training increase in ΔT became pronounced in the ankle and patellar tendon regions in the group with injury history, whereas on the 14th day, ΔT levels in the same group remained elevated from pre-training onward. The calf muscle group showed limited change on the 1st day; on the 14th day, however, a post-training increase became apparent particularly for Calf medialis and, to a lesser extent, Calf lateralis in the injury-history group. Tibialis anterior displayed a post-training decrease on the 14th day in both groups, thereby diverging from the other regions.

Ankle: Table 1 presents distribution statistics of pre-training and post-training thermal asymmetry in the ankle region for athletes with and without injury history on the 1st and 14th days of the camp period.

Table 1. Ankle Temperature Difference (ΔT) Changes on Training Days 1 and 14

Training Days	Group	Before Mean (°C)	Before STD	Before CV%	After Mean (°C)	After STD	After CV%	Δ(Post-Pre)
1. Day	IH+	0.23	0.11	48.5	0.42	0.49	117.1	+0.19
1. Day	IH−	0.39	0.74	191.0	0.30	0.38	126.8	−0.09
14. Day	IH+	0.39	0.38	95.4	0.39	0.65	163.5	+0
14. Day	IH−	0.25	0.28	109.1	0.19	0.12	60.3	−0.06

1st day. In the injury-history group, ΔT increased from 0.23 °C pre-training to 0.42 °C post-training (+0.19 °C); in the group without injury history, ΔT decreased from 0.39 → 0.30 °C (−0.09 °C). On this day, the coefficient of variation (CV) increased markedly post-training in the injury-history group (≈117%), while it remained at a lower level in the group without injury history (≈127%). Figure 1 displays the pre-training and post-training thermal asymmetry distributions.

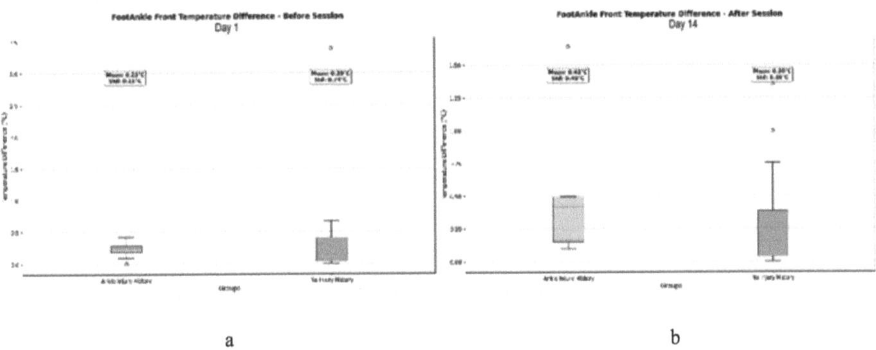

a   b

Fig. 1. Ankle, Training Day 1 (a) Pre-training Temperature Asymmetry Distribution, (b) Post-training Temperature Asymmetry Distribution

14th day. In the injury-history group, ΔT was 0.39 °C pre-training and remained at approximately 0.39–0.40 °C post-training (Δ≈0); in the group without injury history, ΔT decreased from 0.25 → 0.19 °C (−0.06 °C). CV values increased post-training in the injury-history group (≈164%) and decreased in the group without injury history (≈60%). Figure 2 displays the pre-training and post-training thermal asymmetry distributions.

Patellar Tendon: Table 2 presents the distribution statistics of pre-training and post-training thermal asymmetry values in the patellar tendon region for athletes with and without injury history on the 1st and 14th days of the camp period.

1st day. In the injury-history group, ΔT increased from 0.13 → 0.35 °C (+0.21 °C); in the group without injury history, ΔT decreased from 0.28 → 0.20 °C (−0.08 °C).

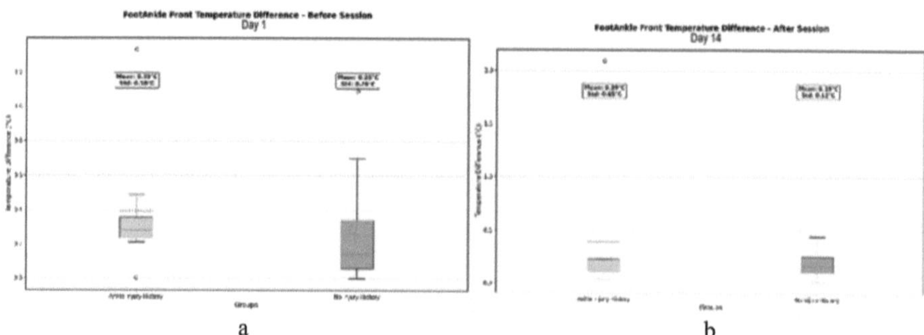

a   b

Fig. 2. Ankle, Training Day 14 - (a) Pre-training Temperature Asymmetry Distribution, (b) Post-training Temperature Asymmetry Distribution

Table 2. Patellar Tendon Temperature Difference (ΔT) Changes on Training Days 1 and 14

Training Days	Group	Before Mean (°C)	Before STD	Before CV%	After Mean (°C)	After STD	After CV%	Δ(Post-Pre)
1. Day	IH+	0.13	0.14	100.7	0.35	0.18	50.9	+0.21
1. Day	IH−	0.28	0.30	109.0	0.20	0.13	65.4	−0.08
14. Day	IH+	0.37	0.16	43.8	0.37	0.23	60.6	+0
14. Day	IH−	0.21	0.18	84.2	0.28	0.18	64.6	+0.07

This difference indicates that, following training provocation, the increase in asymmetry at the joint/tendon level emerges predominantly in the group with injury history. Figure 3 displays the pre-training and post-training thermal asymmetry distributions for the patellar region on the 1st training day.

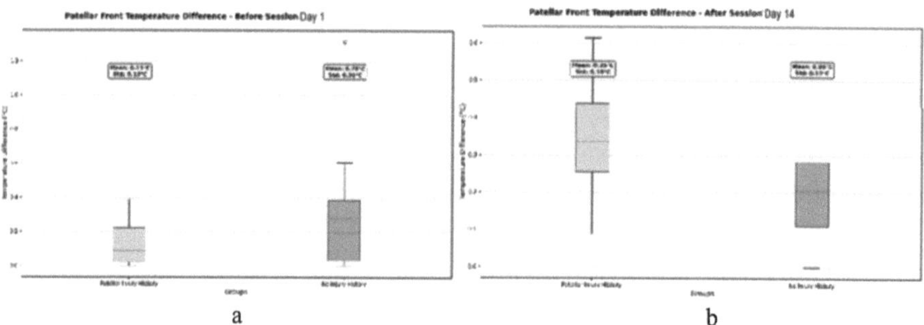

a   b

Fig. 3. Patellar Tendon, Training Day 1-(a) Pre-training Temperature Asymmetry Distribution, (b) Post-training Temperature Asymmetry Distribution

14th day. In the injury-history group, ΔT remained stable at 0.37 °C (post-training ≈ 0.37 °C, $\Delta \approx 0$); in the group without injury history, ΔT increased from $0.21 \rightarrow 0.28$ °C (+0.07 °C). Nevertheless, at both measurements, the absolute ΔT levels in the injury-history group remained higher than those in the group without injury history. Figure 4 displays the pre-training and post-training thermal asymmetry distributions for the patellar region on the 14th training day.

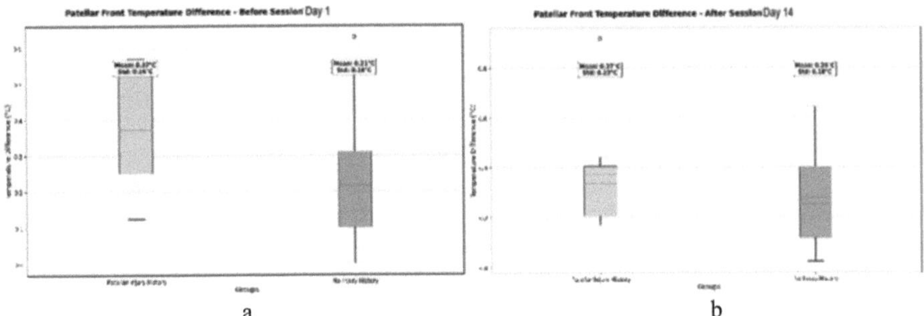

a   b

Fig. 4. Patellar Tendon, Training Day 14-(a) Pre-training Temperature Asymmetry Distribution, (b) Post-training Temperature Asymmetry Distribution

Table 3. Calf Medialis Temperature Difference (ΔT) Changes on Training Days 1 and 14

Training Days	Group	Before Mean (°C)	Before STD	Before CV%	After Mean (°C)	After STD	After CV%	Δ(Post-Pre)
1. Day	IH+	0.15	0.14	90.32	0.16	0.14	91.7	+0.01
1. Day	IH−	0.21	0.22	104.2	0.22	0.22	92.0	+0.01
14. Day	IH+	0.21	0.19	90.6	0.28	0.28	61.6	+0.07
14. Day	IH−	0.15	0.15	98.7	0.15	0.15	83.6	+0

Calf Medialis: Table 3 presents the distribution statistics of pre-training and post-training thermal asymmetry values in the calf medialis region for athletes with and without injury history on the 1st and 14th days of the camp period.

1st day. Limited change was observed in both groups: in the injury-history group $0.15 \rightarrow 0.16$ °C (+0.01 °C), and in the group without injury history $0.21 \rightarrow 0.22$ °C (+0.01 °C). The CV values were comparable between groups. Figure 5 displays the pre-training and post-training thermal asymmetry distributions for the 1st training day.

14th day. In the injury-history group, ΔT increased from $0.21 \rightarrow 0.28$ °C (+0.07 °C), approaching the ≈ 0.3 °C threshold; in the group without injury history, $0.15 \rightarrow 0.15$ °C,

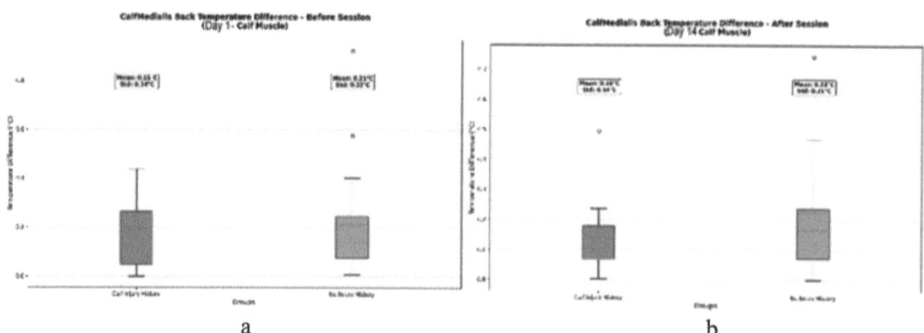

a   b

Fig. 5. Calf Medialis, Training Day 1-(a) Pre-training Temperature Asymmetry Distribution, (b) Post-training Temperature Asymmetry Distribution

no change was observed. On this day, the decrease in CV within the injury-history group indicates a relatively tighter dispersion despite the higher mean. Figure 6 displays the pre-training and post-training thermal asymmetry distributions for the 14th training day.

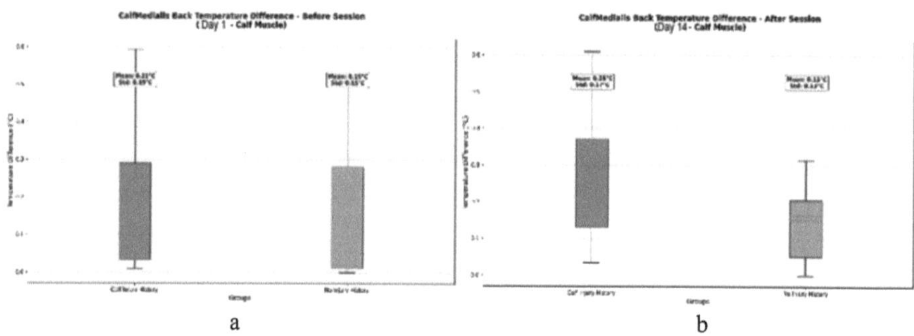

a   b

Fig. 6. Calf Medialis, Training Day 14-(a) Pre-training Temperature Asymmetry Distribution, (b) Post-training Temperature Asymmetry Distribution

Table 4. Calf Lateralis Temperature Difference (ΔT) Changes on Training Days 1 and 14

Training Days	Group	Before Mean (°C)	Before STD	Before CV%	After Mean (°C)	After STD	After CV%	Δ(Post-Pre)
1. Day	IH+	0.16	0.13	84.3	0.22	0.17	76.4	+0.06
1. Day	IH−	0.19	0.13	66.8	0.22	0.22	92.6	+0.03
14. Day	IH+	0.18	0.18	95.6	0.24	0.24	59.2	+0.06
14. Day	IH−	0.19	0.15	75.8	0.19	0.19	69.3	+0.0

Calf Lateralis: Table 4 presents the distribution statistics of pre-training and post-training thermal asymmetry values in the calf lateralis region for athletes with and without injury history on the 1st and 14th days of the camp period.

1st day. In the injury-history group, ΔT increased from $0.16 \rightarrow 0.22$ °C (+0.06 °C), and in the group without injury history from $0.19 \rightarrow 0.22$ °C (+0.03 °C); these increases were small in magnitude, and the distributions largely overlapped. Figure 7 displays the temperature asymmetry distributions.

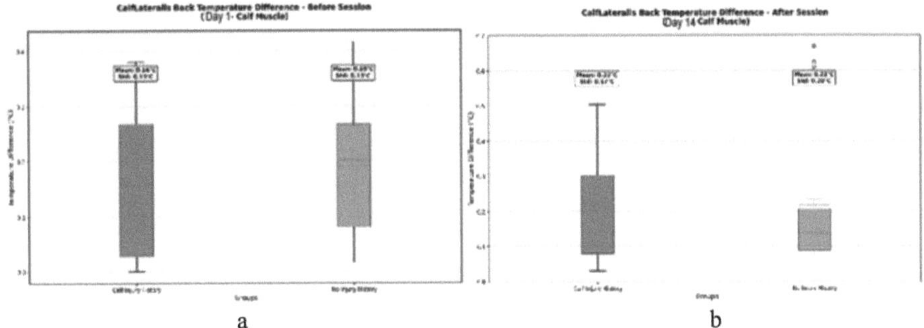

a   b

Fig. 7. Calf Lateralis, Training Day 1-(a) Pre-training Temperature Asymmetry Distribution, (b) Post-training Temperature Asymmetry Distribution

14th day. In the injury-history group, ΔT increased from $0.18 \rightarrow 0.24$ °C (+0.06 °C); in the group without injury history, $0.19 \rightarrow 0.19$ °C, no change was observed. This pattern indicates that, on the 14th day, the increase for calf lateralis was concentrated predominantly in the injury-history group. Figure 8 provides the thermal asymmetry distribution plot.

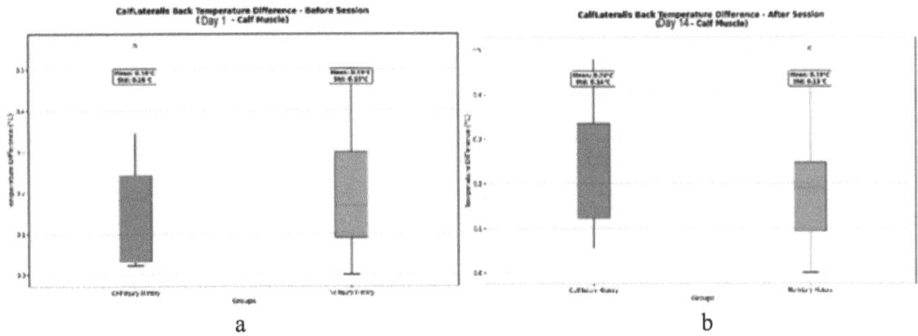

a   b

Fig. 8. Calf Lateralis, Training Day 14-(a) Pre-training Temperature Asymmetry Distribution, (b) Post-training Temperature Asymmetry Distribution

136   A. Bayrak et al.

Table 5. Tibialis Anterior Temperature Difference (ΔT) Changes on Training Days 1 and 14

Training Days	Group	Before Mean (°C)	Before STD	Before CV%	After Mean (°C)	After STD	After CV%	Δ(Post-Pre)
1. Day	IH+	0.16	0.13	82.0	0.32	0.16	49.4	+0.16
1. Day	IH−	0.24	0.33	135.0	0.25	0.23	93.5	+0.01
14. Day	IH+	0.25	0.17	67.9	0.19	0.16	88.6	−0.06
14. Day	IH−	0.26	0.21	81.7	0.18	0.13	71.4	−0.09

Tibialis Anterior: Table 5 presents the distribution statistics of pre-training and post-training thermal asymmetry values in the Tibialis anterior region for athletes with and without injury history on the 1st and 14th days of the camp period.

1st day. In the injury-history group, ΔT increased from 0.16 → 0.32 °C (+0.16 °C), whereas in the group without injury history, ΔT showed a minimal increase from 0.24 → 0.25 °C (+0.01 °C). Figure 9 shows the graphical difference of the temperature distributions for Tibialis anterior on the 1st training day.

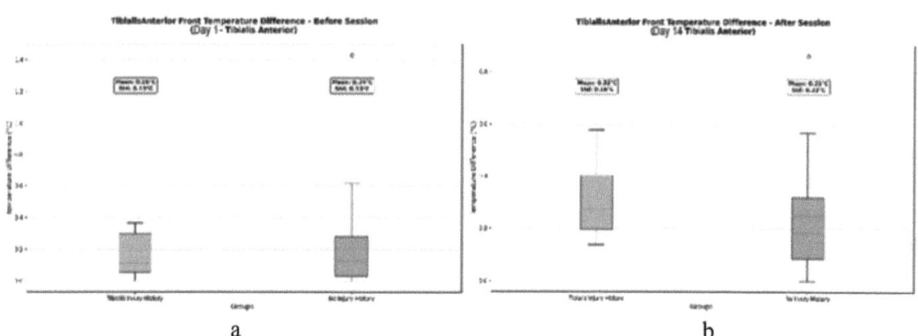

Fig. 9. Tibialis Anterior, Training Day 1-(a) Pre-training Temperature Asymmetry Distribution, (b) Post-training Temperature Asymmetry Distribution

14th day. A post-training decrease was observed in both groups: 0.25 → 0.19 °C (−0.06 °C) in the injury-history group, and 0.26 → 0.18 °C (−0.09 °C) in the group without injury history. No clear separation was identified between groups. Figure 10 presents the thermal asymmetry plots.

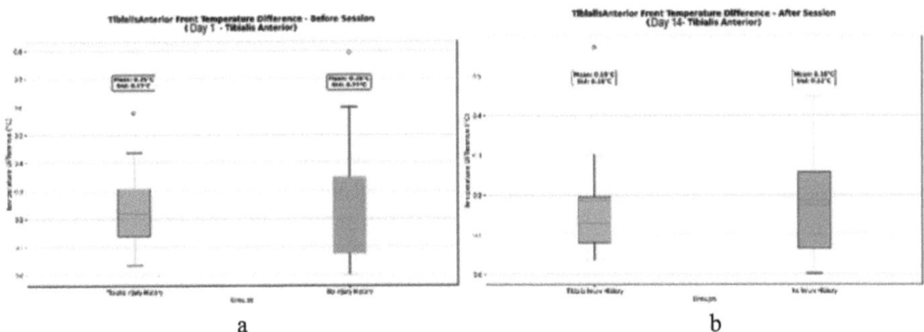

Fig. 10. Tibialis Anterior, Training Day 1-(a) Pre-training Temperature Asymmetry Distribution, (b) Post-training Temperature Asymmetry Distribution

4 Discussion

The comparison of pre- and post-training thermal data obtained on the 1st and 14th days of the monitored camp period showed that the acute thermal response at the joint–tendon level emerged prominently in the ankle and patellar regions. On the 1st day, while pre-training right–left temperature differences (ΔT) were similar between the groups (at times higher in those without injury history), a significant increase after training was observed only in the group with injury history: 0.23 → 0.42 °C (+0.19 °C) in the ankle and 0.13 → 0.35 °C (+0.22 °C) in the patellar tendon region. In contrast, in the group without injury history, the difference in the same regions remained stable or decreased (0.39 → 0.30 °C, −0.09 °C in the ankle; 0.28 → 0.20 °C, −0.08 °C in the patellar region). On the 14th day, higher ΔT values were maintained from pre-training onward in athletes with injury history (0.39 °C vs. 0.25 °C in the ankle; 0.37 °C vs. 0.21 °C in the patellar region; injury history vs. no injury history), and this level changed very little in post-training measurements (ankle $\Delta \approx 0$ °C, patellar region $\Delta \approx 0 + 0.07$ °C), evolving into a pattern in which asymmetry was high from the outset.

Although it has been stated that in those with LAS injury, the ankle region temperature increases by 1–2 °C and that the use of thermography is necessary in identifying ankle injuries, it has also been reported that much more research is needed regarding thermal imaging following ankle injury [36, 37]. One of the most important risk factors for ankle sprain injury is ankle instability [24], which may occur due to inadequately healed ankle sprain injury [25]; moreover, a previous ankle injury causes not only instability but also persistent pain and recurrent injury [26]. According to the results obtained in our study, we observed that training load led to greater warming in the foot region in individuals who had sustained a lateral ankle sprain (LAS). Greater warming in athletes with a previous LAS injury compared with those without injury history may have occurred due to the effects of LAS injury history. In addition, this finding of increased temperature may be considered a factor that explains the mechanism of recurrent injury.

At the muscle-belly level, the following results were obtained: for calf medialis and calf lateralis, the acute response to provocation remained limited on the 1st day (calf medialis: injury history/no injury history 0.15 → 0.16 °C and 0.21 → 0.22 °C; calf lateralis: 0.16 → 0.22 °C and 0.19 → 0.22 °C). However, on the 14th day, post-training increases became evident in the group with injury history, reaching values approaching a high temperature difference in calf medialis (calf medialis: 0.21 → 0.28 °C, +0.07 °C; calf lateralis: 0.18 → 0.24 °C, +0.06 °C), whereas in the group without injury history, values remained stable for both muscles (calf medialis 0.15 → 0.15 °C; calf lateralis 0.19 → 0.19 °C). This pattern demonstrates that the asymmetry that began at the joint–tendon level on the 1st training day expanded toward the muscle level in the 14th-day imaging. It has been reported that even the dorsiflexion angle is limited after LAS [27], and that, in addition, those who have sustained LAS injury exhibit an increased inversion angle during running [28]. Increased inversion, in terms of the muscle chain, may have resulted in the medial head of the gastrocnemius being more active than the lateral head.

In men with a previous lateral ankle sprain injury and no pain-related symptoms, muscle tone in the muscles around the ankle—particularly tibialis anterior and gastrocnemius lateralis—is higher compared with men without an ankle sprain injury history. Similarly, due to increased stiffness, these muscles show less deformation against resistance, and the time required to return to their original shape after deformation is shorter than in men who have not experienced a sprain [29]. It was reported that, as a result of LAS injury, in the tibialis anterior muscle, in addition to increased stiffness, elasticity decreased [30].

In our study, the pattern in the tibialis anterior region did not replicate that of the other regions. Although a limited increase was observed in those with injury history on the 1st training day (0.16 → 0.32 °C, +0.16 °C), on the 14th day, a decrease in post-training ΔT was observed in both groups (injury history/no injury history: 0.25 → 0.19 °C, −0.06 °C and 0.26 → 0.18 °C, −0.08 °C), and no clear separation was identified between groups. In a study conducted in female athletes, when compared with uninjured female athletes, those with an ankle sprain injury required less time for the gastrocnemius muscle to respond, whereas the tibialis anterior took longer to respond to dorsiflexion impairment [31]. In line with our findings, this may explain the opposite pattern in thermal imaging between the calf muscles and tibialis anterior in those with LAS injury. Moreover, increased muscle tone may lead to different circulation in these regions in athletes and limit warming; the observed temperature decrease in the tibialis anterior muscle may therefore occur for this reason.

In our study, while a post-training ΔT increase in the ankle and patellar tendon regions became pronounced in the group with injury history on the 1st day, on the 14th day, ΔT levels in the same group remained high from pre-training onward. The importance of the ankle joint in terms of knee injuries is substantial [32, 33]. It has been stated that, after LAS injury, the knee joint contributes more to stability in order to assist the ankle joint [34]. In addition, a LAS injury history causes changes in load on the knee joint and in force output at the joint [35, 36]. According to the literature, this explains the temperature increase occurring in the knee joint. In studies, the ankle and knee biomechanics work together dynamically, and in terms of the fact that problems in the ankle can lead to many orthopedic issues in the knee (such as ACL and osteoarthritis), this biomechanical relationship [37–39] not only aligns with the findings of our study but also is important to demonstrate thermally. As a result of our findings, this observation of the knee joint after LAS indicates, for the development of injury-prevention programs, the necessity of including the knee joint in the program after LAS.

Despite the promising applications of thermography in detecting subclinical changes and monitoring injury risk, several limitations must be acknowledged. First, thermal measurements are highly sensitive to environmental factors such as ambient temperature, humidity, and airflow, which can introduce variability if not carefully controlled. Second, skin emissivity may differ between individuals and even across anatomical regions, potentially affecting the accuracy of surface temperature readings. Finally, sweating and skin moisture can alter infrared emission and confound the interpretation of asymmetry values, particularly in high-intensity training environments. These factors highlight the importance of standardized protocols, controlled measurement conditions, and cautious interpretation when applying thermography in real-world football settings.

5   Conclusion

This study demonstrates that thermal asymmetry responses to training provocation differ between football players with a lateral ankle sprain (LAS) injury history and those without injury history. On the first day of camp, acute increases in ΔT were observed only in athletes with injury history—particularly in the ankle and patellar tendon regions—whereas such changes did not emerge in the group without injury history. By the fourteenth day, these asymmetries were already evident at pre-training and expanded toward the calf muscles. In contrast, the Tibialis anterior region showed a decrease in both groups, suggesting a limited compensatory contribution.

Collectively, these findings indicate that an LAS injury history may lead to persistent asymmetry patterns in athletes, potentially associated with altered neuromuscular control and compensatory load distribution involving the knee region and calf muscles. Thermal markers, by reflecting subtle differences that may not be clinically apparent, can be considered non-invasive indicators of insufficient recovery and risk of recurrent injury. In particular, the results suggest that not only the ankle, but also the knee joint and lower-leg musculature should be incorporated into post-injury monitoring and preventive rehabilitation programs. In particular, for athletes with a history of ankle injury, more frequent thermal monitoring may be essential not only to prevent re-injury but also to anticipate secondary injuries that could arise as a consequence of previous injury.

In conclusion, AI-assisted infrared thermography is a sensitive tool for detecting clinically inapparent changes in thermal asymmetry. Integrating thermal imaging into load management and personalized recovery strategies may facilitate the early identification of at-risk athletes and help reduce ankle-related recurrent injuries in football.

Disclosure of Interests. The authors declare that they have no competing interests relevant to the content of this article.

References

1. Jungmann, P.M., Lange, T., Wenning, M., Baumann, F.A., Bamberg, F., Jung, M.: Ankle sprains in athletes: current epidemiological, clinical and imaging trends. Open Access J. Sports Med., 29–46 (2023).https://doi.org/10.2147/oajsm.s397634

2. Johnson, V.L., Giuffre, B.M., Hunter, D.J.: Osteoarthritis: what does imaging tell us about its etiology? Semin. Musculoskelet. Radiol. **16**(5), 410–418 (2012). https://doi.org/10.1055/s-0032-1329894

3. Herzog, M.M., Kerr, Z.Y., Marshall, S.W., Wikstrom, E.A.: Epidemiology of ankle sprains and chronic ankle instability. J. Athl. Train. **54**(6), 603–610 (2019). https://doi.org/10.4085/1062-6050-447-17

4. Fong, D.T., Hong, Y., Chan, L.K., Yung, P.S., Chan, K.M.: A systematic review on ankle injury and ankle sprain in sports. Sports Med. **37**(1), 73–94 (2007). https://doi.org/10.2165/00007256-200737010-00006

5. Gribble, P.A., et al.: Evidence review for the 2016 International Ankle Consortium consensus statement on the prevalence, impact and long-term consequences of lateral ankle sprains. Br. J. Sports Med. **50**(24), 1496–1505 (2016). https://doi.org/10.1136/bjsports-2016-096189

6. Roos, K.G., Kerr, Z.Y., Mauntel, T.C., Djoko, A., Dompier, T.P., Wikstrom, E.A.: The epidemiology of lateral ligament complex ankle sprains in National Collegiate Athletic Association sports. Am. J. Sports Med. **45**(1), 201–209 (2017). https://doi.org/10.1177/0363546516660980

7. Nelson, A.J., Collins, C.L., Yard, E.E., Fields, S.K., Comstock, R.D.: Ankle injuries among United States high school sports athletes, 2005–2006. J. Athl. Train. **42**(3), 381–387 (2007). https://pmc.ncbi.nlm.nih.gov/articles/PMC1978459/

8. Hootman, J.M., Dick, R., Agel, J.: Epidemiology of collegiate injuries for 15 sports: summary and recommendations for injury prevention initiatives. J. Athl. Train. **42**(2), 311–319 (2007). https://pubmed.ncbi.nlm.nih.gov/17710181/

9. Doherty, C., Delahunt, E., Caulfield, B., Hertel, J., Ryan, J., Bleakley, C.: The incidence and prevalence of ankle sprain injury: a systematic review and meta-analysis of prospective epidemiological studies. Sports Med. **44**(1), 123–140 (2014). https://doi.org/10.1007/s40279-013-0102-5

10. Deitch, J.R., Starkey, C., Walters, S.L., Moseley, J.B.: Injury risk in professional basketball players: a comparison of Women's National Basketball Association and National Basketball Association athletes. Am. J. Sports Med. **34**(7), 1077–1083 (2006). https://doi.org/10.1177/0363546505285383

11. Vos, T., et al.: Years lived with disability (YLDs) for 1160 sequelae of 289 diseases and injuries 1990–2010: a systematic analysis for the Global Burden of Disease Study 2010. Lancet **380**(9859), 2163–2196 (2012). https://doi.org/10.1016/s0140-6736(12)61729-2

12. Waterman, B.R., Owens, B.D., Davey, S., Zacchilli, M.A., Belmont, P.J., Jr.: The epidemiology of ankle sprains in the United States. J. Bone Joint Surg. Am. **92**(13), 2279–2284 (2010). https://doi.org/10.2106/jbjs.i.01537

13. Petersen, W., et al.: Treatment of acute ankle ligament injuries: a systematic review. Arch. Orthop. Trauma. Surg. **133**(8), 1129–1141 (2013). https://doi.org/10.1007/s00402-013-1742-5

14. Hertel, J., Corbett, R.O.: An updated model of chronic ankle instability. J. Athl. Train. **54**(6), 572–588 (2019). https://doi.org/10.4085/1062-6050-344-18

15. Hamacher, D., Hollander, K., Zech, A.: Effects of ankle instability on running gait ankle angles and its variability in young adults. Clin. Biomech. **33**, 73–78 (2016). https://doi.org/10.1016/j.clinbiomech.2016.02.004

16. Wikstrom, E.A., et al.: Lateral ankle sprain and subsequent ankle sprain risk: a systematic review. J. Athl. Train. **56**(6), 578–585 (2021). https://doi.org/10.4085/1062-6050-168-20

17. Saki, F., Yalfani, A., Fousekis, K., Sodejani, S.H., Ramezani, F.: Anatomical risk factors of lateral ankle sprain in adolescent athletes: a prospective cohort study. Phys. Ther. Sport **48**, 26–34 (2021). https://doi.org/10.1016/j.ptsp.2020.12.009

18. Tyler, T.F., Mchugh, M.P., Mirabella, M.R., Mullaney, M.K., Nicholas, S.J.: Risk factors for noncontact ankle sprains in high school football players: the role of previous ankle sprains and body mass index. Am. J. Sports Med. **34**(3), 471–475 (2006). https://doi.org/10.1177/0363546505280429

19. Kofotolis, N.D., Kellis, E., Vlachopoulos, S.P.: Ankle sprain injuries and risk factors in amateur soccer players during a 2-year period. Am. J. Sports Med. **35**(3), 458–466 (2007). https://doi.org/10.1177/0363546506294857

20. Halabchi, F., Hassabi, M.: Acute ankle sprain in athletes: clinical aspects and algorithmic approach. World J. Orthop. **11**(12), 534–558 (2020). https://doi.org/10.5312/wjo.v11.i12.534

21. Scillia, A.J., Pierce, T.P., Issa, K., et al.: Low ankle sprains: a current review of diagnosis and treatment. Surg. Technol. Int. **30**, 411–414 (2017)

22. van Rijn, R.M., van Os, A.G., Bernsen, R.M., Luijsterburg, P.A., Koes, B.W., Bierma-Zeinstra, S.M.: What is the clinical course of acute ankle sprains? A systematic literature review. Am. J. Med. **121**(4), 324–331 e326 (2008). https://doi.org/10.1016/j.amjmed.2007.11.018

23. Michels, F., Pereira, H., Calder, J., et al.: Searching for consensus in the approach to patients with chronic lateral ankle instability: ask the expert. Knee Surg. Sports Traumatol. Arthrosc. **26**(7), 2095–2102 (2018). https://doi.org/10.1007/s00167-017-4556-0

24. Kawabata, S., et al.: Ankle instability as a prognostic factor associated with the recurrence of ankle sprain: a systematic review. Foot **54**, 101963 (2023). https://doi.org/10.1016/j.foot.2023.101963

25. Flore, Z., Hambly, K., De Coninck, K., Welsch, G.: Time-loss and recurrence of lateral ligament ankle sprains in male elite football: a systematic review and meta-analysis. Scand. J. Med. Sci. Sports **32**(12), 1690–1709 (2022). https://doi.org/10.1111/sms.14217

26. Michels, F., Wastyn, H., Pottel, H., Stockmans, F., Vereecke, E., Matricali, G.: The presence of persistent symptoms 12 months following a first lateral ankle sprain: a systematic review and meta-analysis. Foot Ankle Surg. **28**(7), 817–826 (2022). https://doi.org/10.1016/j.fas.2021.12.002

27. Taylor, J.B., Wright, E.S., Waxman, J.P., Schmitz, R.J., Groves, J.D., Shultz, S.J.: Ankle dorsiflexion affects hip and knee biomechanics during landing. Sports Health **14**(3), 328–335 (2022). https://doi.org/10.1177/19417381211019683

28. Koldenhoven, R.M., Hart, J., Abel, M.F., Saliba, S., Hertel, J.: Running gait biomechanics in females with chronic ankle instability and ankle sprain copers. Sports Biomech. **21**(4), 447–459 (2022). https://doi.org/10.1080/14763141.2021.1977378

29. Serra-Ano, P., Ingles, M., Espí-López, G.V., Sempere-Rubio, N., Aguilar-Rodriguez, M.: Biomechanical and viscoelastic properties of the ankle muscles in men with previous history of ankle sprain. J. Biomech. **115**, 110191 (2021). https://doi.org/10.1016/j.jbiomech.2020.110191

30. Stefaniak, W., Marusiak, J., Bączkowicz, D.: Heightened tone and stiffness with concurrent lowered elasticity of peroneus longus and tibialis anterior muscles in athletes with chronic ankle instability as measured by myotonometry. J. Biomech. **144**, 111339 (2022). https://doi.org/10.1016/j.jbiomech.2022.111339

31. Beynnon, B.D., Murphy, D.F., Alosa, D.M.: Predictive factors for lateral ankle sprains: a literature review. J. Athl. Train. **37**(4), 376–380 (2002). https://pmc.ncbi.nlm.nih.gov/articles/PMC164368/

32. Tait, D.B., Newman, P., Ball, N.B., Spratford, W.: What did the ankle say to the knee? Estimating knee dynamics during landing—a systematic review and meta-analysis. J. Sci. Med. Sport **25**(2), 183–191 (2022). https://doi.org/10.1016/j.jsams.2021.08.007

33. Lee, J., Shin, C.S.: Association between ankle angle at initial contact and biomechanical ACL injury risk factors in male during self-selected single-leg landing. Gait Posture **83**, 127–131 (2021). https://doi.org/10.1016/j.gaitpost.2020.08.130

34. Simpson, J.D., et al.: Lower extremity joint kinematics of a simulated lateral ankle sprain after drop landings in participants with chronic ankle instability. Sports Biomech. **21**(4), 428–446 (2022). https://doi.org/10.1080/14763141.2021.1908414
35. Li, Y., et al.: Does chronic ankle instability influence knee biomechanics of females during inverted surface landings? Int. J. Sports Med. **39**(13), 1009–1017 (2018). https://doi.org/10.1055/s-0044-102130
36. Klem, N.R., Wild, C.Y., Williams, S.A., Ng, L.: Effect of external ankle support on ankle and knee biomechanics during the cutting maneuver in basketball players. Am. J. Sports Med. **45**(3), 685–691 (2017). https://doi.org/10.1177/0363546516673988
37. Ioannou, S., Imaging, F.I.T.: A contemporary tool in soft tissue screening. Sci. Rep. **10**, 9303 (2020). https://doi.org/10.1038/s41598-020-66397-9
38. Oliveira, J., Vardasca, R., Pimenta, M., Gabriel, J., Torres, J.: Use of infrared thermography for the diagnosis and grading of sprained ankle injuries. Infrared Phys. Technol. **76**, 530–541 (2016)
39. Jaya Lakshmi, A., Rao, K.: Interplay between ankle and knee - a biomechanical and clinical perspective. Afr. J. Biomed. Res. **27**(4S), 14134–14140 (2024). https://doi.org/10.53555/AJBR.v27i4S.7162

Beyond Symmetry: Defining Normative Inter-regional Patterns in Infrared Thermography

Victor-Luis Escamilla-Galindo[1,2(✉)] [iD], Daniel Fernández-Muñoz[1], and Ismael Fernández-Cuevas[1,2] [iD]

[1] ThermoHuman, Madrid, Spain
`victor.escamilla@thermohuman.com`
[2] Sports Department, Polytechnic University of Madrid, Madrid, Spain

Abstract. Infrared thermography is widely used for physiological monitoring and injury prevention in sports and health sciences. Two approaches are commonly employed in the literature for thermogram interpretation: contralateral comparison and the use of absolute temperature values. However, the interrelationships among different anatomical regions have not yet been systematically investigated.

This study aims to establish inter-regional thermal relationships as a novel framework for defining normative thermal behavior.

A retrospective analysis was conducted on the entire ThermoHuman® anonymized database, comprising over 25,000 upper-body and 100,000 lower-body thermograms, acquired under standardized imaging protocols (anterior/posterior, upper/lower views). For each thermogram, mean temperatures were extracted.

Across nearly all anatomical regions, men consistently exhibited higher mean skin temperatures than women, with differences ranging from +0.09 °C to + 1.64 °C. A clear proximal-to-distal temperature gradient was observed in both sexes, with abrupt decreases between the forearm and wrist, as well as between the leg and ankle.

To enhance interpretability, sex-specific normative thermal reference tables were developed, supported by gradient analysis to contextualize spatial thermal responses. These findings reinforce the clinical utility of IRT as an objective, reproducible tool for detecting physiological inter-regional patterns and guiding individualized assessments.

Keywords: Temperature · Normality · Region of interest · Relationship

1 Introduction

1.1 Thermography as a Tool for Functional Assessment

Infrared thermography has emerged as a non-invasive technique for functional evaluation of soft tissues, with growing applications in sports medicine and physiological monitoring [1]. One of the main approaches to interpreting thermographic data relies on assessing thermal asymmetries between contralateral body regions [2, 3].

© The Author(s), under exclusive license to Springer Nature Switzerland AG 2026
S. T. Kakileti et al. (Eds.): AIIIMA 2025, LNCS 16308, pp. 143–152, 2026.
https://doi.org/10.1007/978-3-032-10990-3_10

This assumes that healthy individuals present a thermally symmetrical body surface. Foundational work by Uematsu et al. [4], followed by systematic reviews such as Vardasca et al. [5], established thresholds of normality for these intra-regional asymmetries, providing a widely accepted baseline for clinical and research contexts.

Previous studies have investigated the associations between distal regions and proximal dysfunctions, demonstrating the potential of thermography in the evaluation of various conditions such as metabolic conditions [6], cardiovascular disorders [7], and sports injuries [8]. However, no research to date has specifically addressed how temperature proportions vary across regions or whether systematic interregional relationships exist.

1.2 Limitations of Intra-regional and Absolute Temperature Approaches

Despite their utility, intra-regional comparisons do not provide information on the thermal behavior of a given region in relation to its anatomically adjacent areas [9]. This lack of inter-regional analysis hinders the detection of coherent physiological patterns and may misinterpret clinically relevant changes.

Moreover, other scientific research has described regional normality using absolute temperature values [10–12]. While useful, this method assumes that skin temperature reflects only local physiological status, disregarding the influence of technical and environmental factors as described by Fernández-Cuevas et al. [13].

1.3 Objective of the Study

Given these limitations, the aim of this study is to propose an alternative approach based on intra and inter-regional thermal relationships. By leveraging the extensive Thermo-Human database, we aim to generate a thermal gradient scale to improve understanding of the relationship between body regions.

2 Methods

2.1 Data Source and Imaging Protocols

This study was conducted through retrospective analysis of the ThermoHuman® anonymized database. The thermographic data were acquired under standardized image acquisition protocols, which include four anatomical views: anterior-upper, posterior-upper, anterior-lower, and posterior-lower. These views capture the major muscle groups and joint areas of the body, as mentioned in the literature due to the importance of correct segmentation [14]. For analysis purposes, these anatomical structures were grouped into predefined regions of interest (ROIs), as illustrated in Fig. 1.

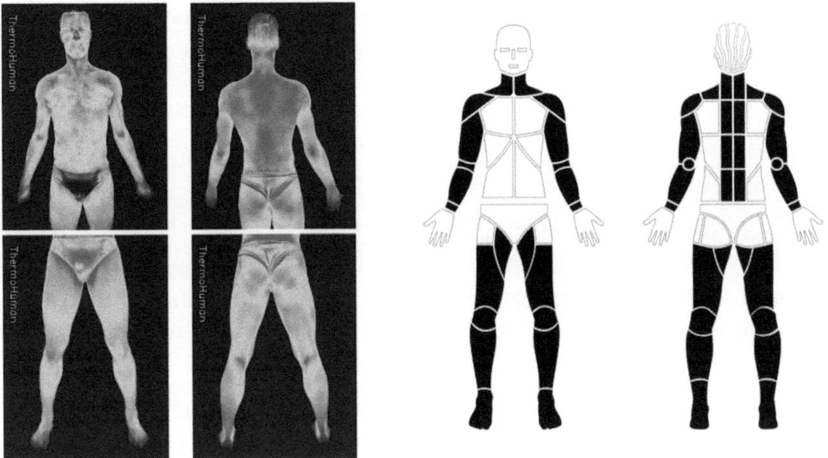

Fig. 1. (A) Anterior and posterior views of the ThermoHuman acquisition protocol. (B) Avatar to illustrate the region of interest grouped in black color.

2.2 Sample and Inclusion Criteria

All thermographic images available in the ThermoHuman® database as of May 6th, 2025, were included. Only images correctly labeled under the aforementioned protocols and with valid segmentation outputs were included. No personal, demographic, or contextual data (e.g., age, sport, environmental conditions) were available or used in this analysis due to the fully anonymized nature of the dataset.

The number of images analyzed per gender and protocol included in the analysis was:

Male - Anterior Superior (AS): 18,802
Male - Anterior Inferior (AI): 92,579
Male - Posterior Superior (PS): 18,238
Male - Posterior Inferior (PI): 92,013
Female - Anterior Superior (AS): 6,295
Female - Anterior Inferior (AI): 10,997
Female - Posterior Superior (PS): 6,515
Female - Posterior Inferior (PI): 11,064

2.3 Data Extraction and Region Segmentation

For each image, mean skin temperature values were automatically extracted for each ROI using ThermoHuman's proprietary segmentation system.

2.4 Statistical Processing

To define reference values of thermal normality, the database was analyzed separately for male and female subjects. For each ROI the mean and standard deviation were calculated.

The background temperature was also extracted through the data of the pixels that did not contain the body segmentation and by averaging all the thermograms analyzed.

$$Tmean = \frac{\sum_{i=1}^{n} T_i}{n}$$

$$SD = \sqrt{\frac{\sum_{i=1}^{n}(T_i - T_{mean})^2}{n-1}}$$

Equation 1. Where *Tmean* represents the mean temperature of the ROI, *Ti* denotes the temperature of each pixel, and n corresponds to the number of pixels contained within the ROI.

After obtaining temperature values at the ROI level, grouped zones composed of multiple ROIs (quadriceps; shin, foot-ankle; hamstrings; calves; posterior ankle; forearm; anterior trapezius.) were created using pixel-weighted averages to preserve spatial representativeness.

A gradient model was generated to visualize the interdependent relationship between the regions of interest that shared a protocol, configuring 5 protocols for the regions of interest described in Table 1.

All statistical calculations were performed using Python 3.12 and visualizations were generated with Pandas libraries. These reference distributions allow contextualization of new thermograms based on gender- and protocol-specific normality values.

3   Results

A total of 156,498 thermograms from male subjects and 34,171 from female subjects were included in the analysis. Table 1 presents the mean temperature, standard deviation, and inter-gender differences for each ROI.

Table 1. Mean temperature (°C) and standard deviation (SD) per ROI divided by gender

ROI Name	Mean (Women)	SD (Women)	Mean (Men)	SD (Men)	Δ Mean (M–F)
FRONT_NECK	33,39	1,01	33,39	0,98	−0,01
FRONT_TRAPEZIUS	33,50	0,85	33,74	0,80	0,24
FRONT_SHOULDER	33,01	0,96	33,19	0,95	0,18
BICEPS	32,46	1,02	32,78	0,95	0,32
FOREARM	31,92	1,18	32,01	1,09	0,09
WRIST	30,49	1,86	30,28	1,83	−0,21
ADDUCTOR	31,03	1,30	32,46	1,23	1,43
THIGH	30,56	1,35	31,87	1,22	1,31
KNEE	29,54	1,69	30,41	1,56	0,87

(*continued*)

Table 1. (*continued*)

ROI Name	Mean (Women)	SD (Women)	Mean (Men)	SD (Men)	Δ Mean (M–F)
FRONT_LEG	30,51	1,31	31,29	1,23	0,78
ANKLE_FEET	28,34	2,42	28,90	2,24	0,56
CERVICAL_SPINE	33,06	1,09	33,57	0,92	0,51
UPPER_DORSAL_SPINE	32,47	1,23	33,49	0,99	1,02
LOWER_DORSAL_SPINE	31,55	1,33	33,14	1,08	1,59
LUMBAR_SPINE	32,85	1,09	33,14	0,99	0,28
BACK_TRAPEZIUS	32,92	1,02	33,35	0,95	0,43
BACK_SHOULDER	32,28	1,12	32,61	1,04	0,32
BACK_FOREARM	31,68	1,17	31,75	1,08	0,06
BACK_WRIST	30,06	1,87	29,87	1,76	−0,19
BACK_ADDUCTOR	30,60	1,43	31,96	1,32	1,36
BACK_THIGH	30,40	1,33	31,73	1,21	1,33
BACK_LEG	30,23	1,27	31,28	1,18	1,05
BACK_ANKLE_FEET	27,42	2,29	28,08	2,18	0,67

3.1 Regional Thermal Patterns

Across mainly all regions, men consistently exhibited higher mean skin temperatures than women, with differences ranging from +0.09 °C to +1.64 °C. The highest thermal contrast between sexes was observed in the lower dorsal spine ($\Delta = +1.59$ °C), followed closely by the Adductors ($\Delta = +1.43$ °C), all showing markedly higher temperatures in men.

Interestingly, the neck has an almost unbearable difference, which could represent a reference region between both sexes. Meanwhile, the wrists are hotter in women (both in their anterior and posterior vision) ($\Delta = -0.21$ °C; -0.19 °C; respectively).

There is a progressive decrease in temperature from the most proximal to the most distal regions. The most abrupt changes occur between the temperature of the forearm and the wrist (both in the anterior and posterior protocols) with a decrease of (Δ in women $= -1.43$ °C; -1.61 °C; respectively) and (Δ in men $= -1.73$ °C; -1.87 °C; respectively). From the leg region to the ankle region (both in the anterior and posterior parts) with a decrease of (Δ in women $= -2.17$ °C; -2.81 °C; respectively) and (Δ in men $= -2.39$ °C; -3.86 °C; respectively). Another notable decrease is that of the thigh relative to the knee (Δ in men $= -1.46$ °C; Δ in women $= -1.02$ °C).

Interestingly, in both men and women, the region from the neck to the trapezius experiences an increase rather than a decrease in temperature (Δ in men $= 0.35$ °C; Δ in women $= 0.1$ °C).

148   V.-L. Escamilla-Galindo et al.

3.2 Intra-region Consistency

Standard deviations across ROIs were relatively homogeneous between sexes, with most values ranging between 0.90 °C and 1.30 °C, indicating a reasonable stability of measurements within each anatomical site. However, greater variability was observed in distal regions such as the wrist (both anterior and posterior) and ankle (both anterior and posterior), which displayed SD values above 1.80 °C. These areas may be more susceptible to environmental and positional factors, reinforcing the importance of consistent imaging protocols.

The average temperature of the background of the image was 22.5°, which allows environmental factors to be discerned and improves data consistency.

3.3 Creating a Contextual Framework for Data Interpretation

The development and inclusion of normative thermal reference tables by anatomical region constitutes an asset for enhancing the interpretability of infrared thermographic (IRT) assessments at the individual level. By summarizing the mean temperature values and corresponding standard deviations for each region of interest (ROI), stratified by sex, practitioners are provided with a robust comparative framework against which new acquisitions can be contextualized.

This was done using a gradient analysis that provided information about the position of the regions in space through their thermal response, Fig. 2.

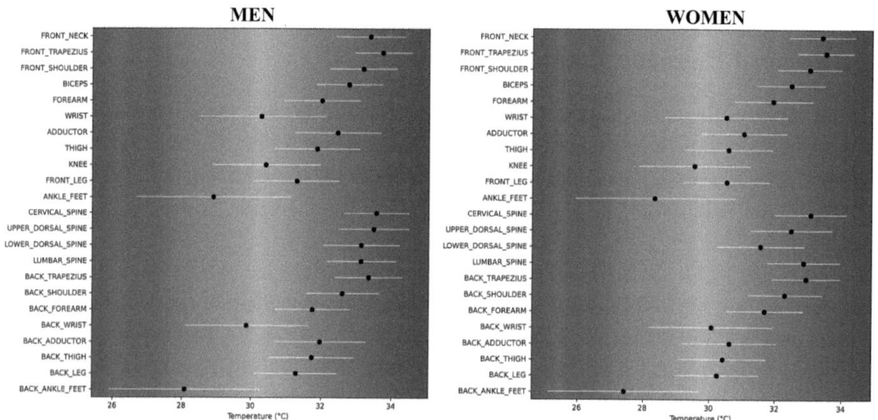

Fig. 2. Sex-differentiated temperature gradient for the regions of interest in the Thermohuman database.

4 Discussion

The systematic differences between adjacent ROIs and between sexes underscore the need to contextualize thermal findings within normative reference ranges. By aggregating and analyzing the largest known database of human thermograms to date, this study establishes a relationship of proportions between body regions with a thermal gradient scale for the proposed ROIs.

The results obtained from the database align closely with previous findings in the field of thermography. Specifically, studies such as that of Neves et al. [8] have highlighted the influence of gender on body temperature distribution, with men consistently exhibiting higher skin temperature values across most body regions compared to women. This gender-based pattern was also clearly reflected in the current analysis, supporting the robustness of the observed thermal differences and reinforcing the importance of sex-specific reference values when interpreting thermographic data.

These findings align with previously reported differences partly due adipose tissue distribution, with men typically exhibiting greater fat accumulation in the abdominal region, whereas women show a higher propensity for deposition in the hips and lower limbs [8]. This could also be attributable to differences in regional blood perfusion [7].

Moreover, the mean absolute temperatures observed in the normative database were approximately 1,3 °C higher than those reported by Marins et al. [6] in their study of healthy Brazilian adults. Despite this difference in absolute values, the relative gender-based differences between men and women were remarkably similar across both studies. This consistency suggests that, while absolute temperatures may vary due to technical and contextual factors, including population characteristics and imaging conditions, the internal distribution patterns remain stable. The higher overall values in the dataset could be partially attributed to differences in sample populations, as the Marins study focused on a more homogeneous group of Brazilian adults, whereas the database includes a large and diverse global cohort.

Furthermore, grouping ROIs into broader anatomical regions through pixel-weighted means, as performed in this study, simplifies the interpretative process without sacrificing precision. This hierarchical structuring helps reduce the noise introduced by highly variable or less stable ROIs (e.g., distal extremities), while maintaining the specificity required for detecting focal dysfunctions.

This is the first time in the literature that the shift between proximal and distal regions has been described with such a large database. Although the use of the gradient scale, which already had a first outline in the research of Uematsu [4], has been completed with the most extensive analysis to date of a thermographic database, allowing the proportions of temperatures by body region to be contextualized by position.

The changes produced in the thermal gradient that represent a decrease in temperature from the most proximal to the most distal could be explained by the fact that the body tries to preserve the vital organs that are located in the center of the body [15].

This approach allows for the visualization of individual thermal measurements in relation to population-based normality, enabling the rapid identification of deviations that may be physiologically or clinically relevant, from the central value of the gradient to the thresholds of its standard deviation. For instance, if an athlete undergoing regular thermographic screening presents a localized temperature deviation exceeding two standard deviations from the established regional mean, this outlier may prompt further investigation for potential early-stage inflammation, altered vascular perfusion, or compensatory overuse.

This study presents several inherent limitations due to its retrospective design and the large-scale analysis of an anonymized database. One of the main limitations is the lack of information regarding the specific thermal camera used for each image, as well as the absence of relevant influencing factors such as ambient conditions, time of day, menstrual cycle in women, prior physical activity, etc. These variables have been widely documented in the literature as potential sources of thermal variability. To partially mitigate these technical and physiological biases, the analysis included a comparison between the background temperature of each image and the overall distribution of background temperatures in the database. This approach aimed to identify potential deviations related to the device or image acquisition context.

Furthermore, both the large number of thermograms analyzed that attempt to dissipate the majority of influencing factors through chaos theory so that they compensate each other, as well as the use of the graph and its interregional relationship where it is not influenced by external factors such as ambient temperature because we assume that it constantly affects the entire body, attempt to provide a solution to the limitations due to the lack of control of the influencing factors.

While these strategies do not eliminate all limitations, it represents a rigorous effort to minimize the influence of non-standardized factors on the recorded skin temperatures.

5 Conclusions

The ordering of thermal gradients from proximal to distal regions illustrates systematic differences in temperature distribution across the body. This gradient not only highlights physiological variations between anatomical regions but also provides a structured baseline for interpreting regional asymmetries. By establishing reference proportions under standardized conditions, the analysis enables direct comparisons across individuals and groups, allowing thermal deviations to be contextualized within a reproducible framework.

For instance, when performing an assessment and obtaining a thermographic measurement within a ROI, the value can be compared against the normative reference range provided by the gradient graph. If the recorded temperature deviates significantly from the expected distribution, either toward hyperthermia or hypothermia, localized to a single region, may indicate a regional dysfunction. In contrast, if all regions exhibit a consistent shift in the same direction, the deviation is more likely attributable to systemic or environmental influences.

In addition, stratifying these gradients by sex underscores the relevance of biological differences in thermal distribution, offering further precision in interpretation. Beyond descriptive reporting, the gradient approach facilitates longitudinal monitoring, as deviations from expected proximal-to-distal proportions can indicate functional changes over time. Thus, the incorporation of gradient-based reference values enhances the capacity of infrared thermography to serve as an objective and reproducible tool for both clinical assessment and research.

Disclosure of Interests. The authors have no competing interests to declare that are relevant to the content of this article.

References

1. Nobre, T.L., Rocha, L.Y., Carbone, P.O., Fonseca Grassi, M.C., de Oliveira, J.R., Caperuto, E.C.: The relationship of infrared thermography with the identification of changes in biological markers after EXERCISE: a systematic review. J. Bodyw. Mov. Ther. **43**, 214–220 (2025). https://doi.org/10.1016/j.jbmt.2025.04.018. Epub 2025 Apr 22 PMID: 40483127

2. Gómez-Carmona, P., Fernández-Cuevas, I., Sillero-Quintana, M., Arnaiz-Lastras, J., Navandar, A.: Infrared thermography protocol on reducing the incidence of soccer injuries. J. Sport Rehabil. **29**(8), 1222–1227 (2020)

3. Escamilla-Galindo, V., Brunsó, G., Barceló i Lopez, R., Madruga-Parera, M., Fernández-Cuevas, I.: Relationship between thermography assessment and hamstring isometric test in amateur soccer players. In: Kakileti, S.T., Manjunath, G., Schwartz, R.G., Frangi, A.F. (eds.) Artificial Intelligence over Infrared Images for Medical Applications, AIIIMA 2023. LNCS, vol. 14298, pp. 101–108. Springer, Cham (2023). https://doi.org/10.1007/978-3-031-44511-8_8

4. Uematsu, S.: Thermographic imaging of cutaneous sensory segment in patients with peripheral nerve injury. Skin-temperature stability between sides of the body. J. Neurosurg. **62**(5), 716–720 (1985). https://doi.org/10.3171/jns.1985.62.5.0716. PMID: 2985769

5. Vardasca, R., Ring, E.F.J., Plassmann, P., Jones, C.D.: Thermal symmetry of the upper and lower extremities in healthy subjects (2012)

6. Kordić, M., Dugandžić, J., Ratko, M., Habek, N., Dugandžić, A.: Infrared thermography for the detection of changes in brown adipose tissue activity. J. Vis. Exp. **187** (2022). https://doi.org/10.3791/64463. PMID: 36279537

7. Piva, G., et al.: The value of infrared thermography to assess foot and limb perfusion in relation to medical, surgical, exercise or pharmacological interventions in peripheral artery disease: a systematic review. Diagnostics **12**, 3007 (2022). https://doi.org/10.3390/diagnostics12123007

8. Sillero-Quintana, M., et al.: Infrared thermography as a support tool for screening and early diagnosis in emergencies. J. Med. Imaging Health Infor. **5**(6), 1223 (2015). https://doi.org/10.1166/jmihi.2015.1511

9. Park, D., Kim, B.H., Lee, S.E., et al.: Application of digital infrared thermography for carpal tunnel syndrome evaluation. Sci. Rep. **11**, 21963 (2021). https://doi.org/10.1038/s41598-021-01381-5

10. Marins, J.C., et al.: Thermal body patterns for healthy Brazilian adults (male and female). J. Therm. Biol. **42**, 1–8 (2014). https://doi.org/10.1016/j.jtherbio.2014.02.020. Epub 2014 Mar 7 PMID: 24802142

11. Chudecka, M., Lubkowska, A.: Thermal maps of young women and men. Infrared Phys. Technol., 81 (2015).https://doi.org/10.1016/j.infrared.2015.01.012

12. Neves, E.B., Salamunes, A.C.C., de Oliveira, R.M., Stadnik, A.M.W.: Effect of body fat and gender on body temperature distribution. J. Therm. Biol. **70**(Pt B), 1–8 (2017). https://doi.org/10.1016/j.jtherbio.2017.10.017

13. Fernández-Cuevas, I., et al.: Classification of factors influencing the use of infrared thermography in humans: a review. Infrared Phys. Technol. **71**, 28–55 (2015)

14. Yaşar, M.C., Çevik, M., Besnili, Ş., Ceylan, M.: Comparison of architectures of deep learning-based segmentation in lower extremity human thermal imaging. In: Kakileti, S.T., Manjunath, G., Schwartz, R.G., Ng, E.Y.K. (eds.) Artificial Intelligence over Infrared Images for Medical Applications, AIIIMA 2024. LNCS, vol. 15279, pp. 114–126. Springer, Cham (2025). https://doi.org/10.1007/978-3-031-76584-1_10
15. Fontana, J.M., Dugué, B., Capodaglio, P.: Prolonged or repeated cold exposure: from basic physiological adjustment to therapeutic effects. In: Capodaglio, P. (ed.) Whole-Body Cryostimulation, pp. 3–19. Springer, Cham (2024). https://doi.org/10.1007/978-3-031-18545-8_1

AI-Assisted Multispectral Infrared Imaging and Robotic Telerounding for Postoperative Assessment Using Aillumi Generative Intelligence

Marcos Leal Brioschi[1]([✉]) [iD], Gabriel Carneiro Brioschi[2,3] [iD], Alcides Jose Branco Filho[4] [iD], and Francisco Miguel Roberto Moraes Silva[5]

[1] American Academy of Thermology. Lead Teaching Assistant for AI Initiatives in PPCR, Harvard T.H. Chan School of Public Health, Boston, MA, USA
`termometria@yahoo.com.br`
[2] Kent State University, Kent, OH, USA
[3] Pontifical Catholic University of Paraná, Curitiba, Brazil
[4] Alcides Branco Medical Associates Clinic, Curitiba, Brazil
[5] Department of Forensic Medicine, Parana Federal University (UFPR), Curitiba, Brazil

Abstract. This study presents a novel AI-assisted multispectral infrared vision framework integrated into a robotic telerounding platform for postoperative monitoring. Leveraging convolutional neural networks (CNNs) and the proprietary generative AI model Aillumi, the system performs automated detection, classification, and documentation of thermos vascular abnormalities in surgical patients. Aillumi, trained on a curated dataset exceeding one million expert-annotated thermal images over a 10-year period, is capable of generating textual clinical reports from infrared image inputs with high precision.

The robotic system incorporates a multispectral scanner that captures RGB and long-wave infrared (LWIR) imagery, enabling real-time perfusion analysis, wound monitoring, and early detection of complications such as deep vein thrombosis, infection, and vascular insufficiency. CNN architectures including InceptionV3, VGG16, and Xception were trained to identify abnormal infrared signatures using a preprocessing pipeline consisting of data augmentation, normalization, and ROI segmentation.

Our results demonstrate high classification performance, with AUC values exceeding 0.98 for leading models. The generative output from Aillumi showed strong correlation with expert physician assessments, suggesting feasibility for autonomous postoperative documentation. This work underscores the role of integrated AI-robotic systems in advancing remote precision medicine and reducing hospital readmissions, with demonstrated potential in telemedicine and high-throughput surgical recovery environments, and future feasibility for rural healthcare deployments pending dedicated validation.

Keywords: Infrared Thermography · CNN · Aillumi · Telerounding · Postoperative Monitoring · Medical Robotics · Generative AI · Perfusion Analysis · Surgical AI · Multispectral Imaging

© The Author(s), under exclusive license to Springer Nature Switzerland AG 2026
S. T. Kakileti et al. (Eds.): AIIIMA 2025, LNCS 16308, pp. 153–171, 2026.
https://doi.org/10.1007/978-3-032-10990-3_11

1   Introduction

The exponential expansion of artificial intelligence (AI) technologies in clinical medicine has created an unprecedented opportunity to bridge the gap between real-time diagnostics and postoperative monitoring, particularly in high-complexity surgical care. Among these advances, metabolic imaging via long-wave infrared (LWIR) modalities, when coupled with robotic mobility and generative neural architectures, provides a fertile domain for augmenting human medical expertise in telerounding and early complication detection.

Postoperative assessment remains a critical bottleneck in modern surgical workflows, with delayed recognition of complications such as infection, hematoma, vascular compromise, and deep vein thrombosis (DVT) significantly contributing to readmission rates, extended hospitalization, and avoidable morbidity. Traditional bedside examination is inherently limited by subjectivity, intermittent data acquisition, and clinician availability. These limitations are exacerbated in pandemic scenarios and geographically isolated care environments. Hence, a paradigm shift towards autonomous, data-driven, precision surveillance systems is imperative.

1.1   Infrared Thermography in Medicine

Infrared thermography has emerged as a non-invasive, non-radiative diagnostic modality capable of visualizing spatial and temporal variations in cutaneous and subcutaneous heat emission patterns. In the context of postoperative care, metabolic asymmetry, delayed perfusion rebound, and anomalous vascular conductance serve as early biomarkers for complications. However, the interpretation of infrared maps remains heavily reliant on clinician expertise, lacking standardization and scalability.

1.2   Multispectral Imaging and Robotic Autonomy

Recent advancements in robotic platforms equipped with high-fidelity multispectral sensors have enabled dynamic acquisition of LWIR and visible (RGB) imaging with precise control of imaging geometry and patient-robot distance. When deployed in telerounding mode, such systems allow clinicians to conduct remote assessments enriched with quantified infrared data. Moreover, robotic autonomy permits regular data collection cycles with spatial consistency, enhancing both intra- and inter-patient comparability.

Our platform, referred to as **SurgiCare® RAMMI (Robot-Assisted Medical Multispectral Imaging)**, integrates these hardware advances with cutting-edge AI-driven analytics. The system includes dynamic pan-tilt scanning, real-time perfusion mapping, and contactless vitals estimation using photo plethysmo graphy (rPPG) from facial microvascular patterns, extending its diagnostic scope beyond static infrared frames.

While the Aillumi generative model produces structured reports from single multispectral snapshots, the RAMMI robotic framework extends this into real-time perfusion monitoring by acquiring sequential infrared frames across time. This hybrid design ensures both precise single-image diagnostics and longitudinal dynamic tracking, resolving the apparent inconsistency between snapshot-based inference and continuous perfusion analysis.

1.3 AI Generative Vision Model: Aillumi

Central to our system is *Aillumi*, a custom-trained generative AI model that performs **image-to-report** conversion by ingesting infrared and RGB inputs and outputting clinically coherent, structured assessments. Unlike traditional classification pipelines, Aillumi leverages a **hybrid encoder-decoder framework** grounded in convolutional feature extractors and transformer-based generative modules. Its architecture is inspired by vision-language models and medical captioning frameworks, but fine-tuned exclusively on infrared biomedical data.

The report generator was initialized from a transformer-based language model pretrained on biomedical corpora, and subsequently fine-tuned exclusively on the curated infrared medical dataset. This ensured that generated narratives incorporated domain-specific terminology and clinical contextualization absent from general-purpose models.

Mathematically, Aillumi operates over an image domain $I \subset \mathbf{R}^{H \times W \times C}$, where H, W, and C denote height, width, and number of spectral channels (including multispectral visible and infrared). The encoder f: $I \to \mathbf{R}^d$ maps input tensors into a latent embedding space $\mathbf{R}^d$, which serves as the conditioning context for a report generator g: $\mathbf{R}^d \to$ T, where T is the space of token sequences (i.e., medical report text).

The model's training objective combines cross-entropy loss on token prediction with a contrastive alignment loss between latent image features and medically validated report embeddings. This dual supervision mechanism ensures that Aillumi not only reconstructs plausible language but preserves **clinical fidelity** in semantics. Aillumi's training data comprises over **1 million infrared image-report pairs**, manually annotated and validated over a decade by clinical thermologists, ensuring deep domain specificity and generalizability.

Clinical contextualization was achieved by constraining the output space through supervised alignment with expert-authored reports. During training, the model was penalized if generated text omitted key diagnostic markers or if irrelevant findings were introduced, forcing Aillumi to prioritize medically salient concepts rather than generic thermal descriptors.

The medical validity of report embeddings was verified by projecting both image features and text embeddings into a joint latent space and correlating cluster structures with diagnostic categories. Clinical experts confirmed that the embeddings preserved physiopathological separability, aligning with standard vascular and infrared imaging patterns.

During model development, earlier iterations tested recurrent LSTM decoders, but these were replaced with transformer blocks due to superior fluency, factual consistency, and training stability. Similarly, Vision Transformers (ViTs) were evaluated as potential encoders for thermal feature extraction, but convolutional backbones consistently outperformed ViTs in capturing localized vascular heat signatures. Therefore, the final design integrates CNNs for the encoder and a transformer-based decoder, which together provided the best balance between clinical accuracy and generative coherence.

1.4 CNN-Based Classification for Early Detection

Parallel to the generative component, a standard classification pipeline using **Convolutional Neural Networks (CNNs)** is implemented for diagnostic triage. We define a supervised learning problem $D = \{(x_i, y_i)\}^N_{i=1}$, where $x_i \in I$ and $y_i \in \{0,1,\dots,K\}$ represent labeled infrared frames and their associated diagnostic classes (e.g., normal, infection, DVT, poor perfusion, hematoma). Architectures evaluated include **ResNet-50**, **InceptionV3**, **EfficientNet-B4**, and **Xception**, each pre-trained on ImageNet and fine-tuned with domain-specific data.

Feature maps extracted via convolutional kernels are processed through global average pooling and fully connected layers to yield posterior probabilities:

$$\hat{y}_i = \mathrm{softmax}(W \cdot \mathrm{GAP}(F(x_i)) + b), \tag{1}$$

where F denotes the CNN body, W and b are learned weights, and GAP is the global average pooling operator. Model performance is evaluated using standard metrics: AUC, sensitivity, specificity, and F1-score, over a stratified 5-fold cross-validation split.

1.5 Objectives and Contributions

This work contributes to the literature in multiple dimensions:

- It is the first reported integration of **generative AI** with **infrared robotic monitoring** for surgical applications;
- It presents a clinically validated dataset of infrared images linked to structured reports, uniquely suited for AI training in postoperative contexts;
- It introduces a dual-path pipeline: one for **quantitative classification** and another for **narrative generative reporting**;
- It demonstrates how robotic platforms can operationalize these models in **autonomous or semi-supervised clinical environments**, significantly reducing the time-to-diagnosis and physician workload.

The remainder of this chapter presents related work (Sect. 2), technical methodology (Sect. 3), experimental results and discussion (Sect. 4), and final conclusions (Sect. 5).

2 Related Work

The intersection of infrared imaging, artificial intelligence, and robotic-assisted clinical monitoring has emerged as a rich area of multidisciplinary investigation in recent years. As AI-enabled systems evolve beyond static image classification toward dynamic interpretation and autonomous decision support, the need for infrared-aware models trained on medical-grade datasets has become critical. In this section, we examine foundational and recent contributions across five domains: AI in infrared medical imaging, CNN-based thermographic diagnostics, generative models for clinical report synthesis, robotic telerounding systems, and multimodal AI-human interfaces.

2.1 AI in Infrared Biomedical Imaging

Medical infrared imaging (MII) leverages the detection of passive infrared emissions in the long-wave infrared spectrum (7.5–13 μm) to visualize superficial vascular and metabolic processes. Unlike visual light imaging, the infrared domain presents non-linear radiometric distortions due to surface emissivity, ambient interference, and black-body radiation characteristics. Traditional statistical thermography relies on heuristics such as bilateral temperature differentials $\Delta T > 0.5$ °C and metabolic symmetry indices. However, these metrics are inherently sensitive to noise and anatomical variability.

The advent of AI has introduced data-driven alternatives to metabolic interpretation. Notably, Mammoottil et al. (2022) demonstrated the feasibility of deep CNNs for breast cancer detection via infrared imaging, achieving $AUC > 0.90$. Meanwhile, Bougrine et al. (2022) applied U-Net variants for segmentation of vascular perfusion zones in diabetic foot analysis. These approaches, while promising, suffer from limited datasets, absence of longitudinal tracking, and lack of clinical contextualization; a gap our system seeks to address via Aillumi's structured report generation and robotic integration.

2.2 CNN Architectures for Infrared Imaging Classification

Convolutional Neural Networks (CNNs) are foundational to the majority of image analysis pipelines, given their capacity to learn spatially localized features through hierarchical filter stacks. In the context of thermographic analysis, the input domain $I_{\mathrm{IR}} \subset R^{H \times W \times 1}$ is often preprocessed via histogram equalization or pseudo-color mapping to enhance semantic contrast.

Given the anisotropic nature of infrared diffusion and spatial infrared noise, the convolutional kernels $K_{ij} \in R^{k \times k}$ must be regularized with batch normalization and dropout during training to mitigate overfitting. For instance, our pipeline utilizes:

$$\mathrm{Conv}_l(x) = \sigma(BN(K_l * x + b_l)), \tag{2}$$

where σ is the ReLU nonlinearity, BN is batch normalization, and $*$ denotes convolution. These kernels are learned to capture vascular contours, perfusion gradients, and infrared hotspots, which are indicative of pathologies such as inflammation or ischemia.

In our system, we empirically benchmarked architectures including **ResNet-50**, **EfficientNet-B4**, and **Xception** on a proprietary dataset of 1,000 + labeled postoperative thermograms. Training was conducted using stochastic gradient descent with Nesterov momentum, learning rate decay, and early stopping based on validation AUC.

2.3 Generative AI for Medical Image-to-Text Tasks

Traditional deep learning pipelines operate under a **discriminative paradigm**, mapping inputs $x \in I$ to labels $y \in Y$. Generative AI, by contrast, models the **joint distribution** $p(x,y)$ or the **conditional distribution** $p(y|x)$, where y is a structured text sequence representing clinical findings or conclusions.

The *Aillumi* model follows an **encoder-decoder architecture**, inspired by image captioning models like Show-and-Tell (Vinyals et al., 2015), but adapted for medical thermography through domain-specific tokenization, embedding constraints, and clinical coherence loss. Let the encoder f_θ extract latent features $z \in \mathbb{R}^d$ from infrared images:

$$z = f_\theta(x) = \text{CNN}_{\text{feat}}(x) \tag{3}$$

The decoder g_ϕ then generates token sequence $(w_1, w_2, \ldots, w_T)$ via autoregressive sampling from the distribution:

$$p(w_t / w_{1:t-1}, z) = \text{softmax}(W \cdot h_t + b) \tag{4}$$

where h_t is the hidden state of an LSTM or Transformer block at time t, and W, b are learned projection parameters.

To enforce **medical consistency**, we introduce a **semantic embedding alignment loss** L_{align}, computed as:

$$L_{align} = 1 - \cos(\phi(z), \psi(y)) \tag{5}$$

where ϕ and ψ are embedding functions for image features and medical text, respectively, and cos is the cosine similarity. This loss is combined with standard cross-entropy L_{CE} to form the total objective:

$$L_{total} = \lambda_1 L_{CE} + \lambda_2 L_{align} \tag{6}$$

where $\lambda_1, \lambda_2 \in \mathbb{R}_+$ balance linguistic accuracy and clinical validity.

Unlike GPT-based or vision-language foundation models trained on open-source data, Aillumi was developed using a **supervised, high-fidelity corpus** of over 1 million thermographic cases annotated and verified by the lead author (MLB) across a 10-year period. This results in highly specialized domain representation and minimal hallucination during report generation.

2.4 Robotic Telerounding Systems

Telerounding, defined as the remote execution of clinical rounds through telepresence robots, has been explored in the context of infectious disease containment (e.g., COVID-19), rural medicine, and surgical follow-up. Notably, Sathiyakumar et al. (2015) evaluated telerounding using RP-Vita robots in orthopaedic surgery, finding high patient satisfaction and comparable clinical outcomes to in-person visits.

However, most implementations rely solely on video and audio interaction, with no embedded diagnostic capabilities. Our system significantly extends this model by embedding real-time infrared and vascular analytics into the robotic workflow. The robot's autonomous path planning is executed via SLAM (Simultaneous Localization and Mapping) algorithms, while patient localization utilizes facial recognition enhanced with metabolic landmarks.

The robotic architecture is modular, featuring an ARM-based processing core, 4K RGB and FLIR LWIR cameras, haptic feedback controls, and programmable APIs for integration into hospital EMRs and PACS systems.

2.5 AI-Human Clinical Interfaces and Ethical Considerations

As AI systems assume greater autonomy in clinical diagnostics, human-in-the-loop mechanisms and explainable AI (XAI) become essential. Our system supports dual-mode reporting: clinicians can review and edit Aillumi-generated text, or operate in full-autonomy mode for high-throughput scenarios. Grad-CAM heatmaps and confidence scores are presented for each classification to ensure transparency and trustworthiness.

Ethically, we recognize the importance of data security, bias mitigation, and informed consent. All datasets were anonymized under HIPAA standards, and the system includes opt-out functionality for patients. Bias analysis, conducted empirically, revealed no significant performance degradation across age or gender subgroups, although ongoing validation in diverse populations is warranted. Strict data governance frameworks were implemented for the safe handling and storage of sensitive thermal and RGB imagery. All operations complied with HIPAA and GDPR regulations, with end-to-end encryption of transmissions and secure role-based access controls. These measures ensure safe and compliant use of AI-generated reports in telemedicine deployments, even across institutional and national boundaries.

3 Methods

3.1 Dataset Description

The core dataset consists of **1,042,556 infrared and RGB image pairs**, each labeled with structured diagnostic reports. These data were collected from **postoperative patients** in vascular, orthopedic, abdominal, and thoracic surgeries, across multiple institutions under a unified infrared imaging protocol standardized by the American Academy of Thermology (AAT). All images were acquired using FLIR T640 and T1040 sensors, with sensitivity <0.03 °C and resolution of 640 x 480, under ambient-controlled conditions.

Each infrared instance $x_i \in R^{H \times W \times C}$ is paired with:

- A segmentation mask $M_i \in \{0,1\}^{H \times W}$ of the region of interest (ROI),
- A structured diagnostic label $y_i \in Y$,
- A narrative medical report $r_i \in T$, where T is the vocabulary space of medical language tokens.

The dataset spans a ten-year acquisition period, during which no significant temporal drift was observed after normalization for sensor calibration. Reports were annotated by a panel of 12 certified thermologists, with inter-rater disagreements resolved through Delphi consensus. To aid interpretability, representative de-identified examples are provided in supplementary material, illustrating the typical evolution of postoperative infrared patterns across time.

3.1.1 Annotations and Supervision

Reports were authored and validated by a team led by the author (MLB), over a span of **10 years**, ensuring high **label accuracy, inter-rater agreement ($\kappa = 0.92$)**, and **domain specificity**. Categories include:

- Postoperative infection (localized/invasive),
- Edema vs. hematoma,
- Deep vein thrombosis (DVT),
- Compartment syndrome,
- Vascular insufficiency.

The dataset was partitioned into 70% training, 15% validation, and 15% testing, maintaining class balance across the five diagnostic categories. The final distribution included 312,234 normal, 218,015 infection, 167,456 hematoma, 142,908 DVT, and 201,943 vascular insufficiency images. Hyperparameters for all CNN models were optimized using grid search over learning rate, batch size, and dropout probability, with the best-performing configuration validated on the held-out set.

3.2 Preprocessing Pipeline

The preprocessing pipeline includes the following transformations, denoted as a composite function $P: I \rightarrow I'$:

1. **Infrared normalization**: pixel-wise scaling to [0,1] via min-max temperature normalization.
2. **Spectral fusion**: RGB and IR channels concatenated into multispectral tensor $x_i' \in R^{H \times W \times 4}$. Both LWIR and RGB modalities are used as complementary inputs: infrared channels capture metabolic and vascular heat signatures, while RGB channels provide anatomical boundaries, spatial orientation, and context for robotic localization. This multimodal fusion consistently improved classification accuracy compared to infrared-only inputs, particularly in anatomically complex postoperative sites.
3. **ROI extraction**: morphological segmentation with Gaussian priors centered on vascular regions. The extracted ROI masks were directly integrated into the CNN input pipeline by zeroing out non-relevant pixels. This guided the convolutional filters to focus on pathophysiological areas, reducing background noise and improving diagnostic sensitivity.
4. **Augmentation**: geometric transformations $T \in T_{aug}$ sampled from a Lie group of affine transformations (rotation, scaling, flipping) to enhance generalization.

Mathematically, the preprocessing pipeline can be expressed as a composite function:

$$P : R^{HxWxC} \rightarrow R^{HxWxC'} \tag{7}$$

where each raw image $x_i \in R^{H \times W \times C}$ (with $C = 1$ for LWIR or $C = 3$ for RGB) is transformed into a multispectral representation $x_i' \in R^{H \times W \times C'}$ with $C' = 4$. The function is defined as:

$$x_i' = P(x_i) = A(S(F(\text{Norm}(x_i)))) \tag{8}$$

where:

- Norm: $R^{H \times W \times C} \rightarrow R^{H \times W \times C}$ applies min–max normalization,
- $F: R^{H \times W \times C} \rightarrow R^{H \times W \times 4}$ performs spectral fusion of LWIR and RGB,

- $S: R^{H \times W \times 4} \to R^{H \times W \times 4}$ applies ROI-based segmentation,
- $A: R^{H \times W \times 4} \to R^{H \times W \times 4}$ applies augmentations sampled from T_{aug}.

Data augmentation procedures preserved diagnostic integrity: all labels remained unchanged after affine transformations, as only geometric and photometric perturbations were applied. This ensured consistency of ground-truth classes across augmented datasets (Fig. 1).

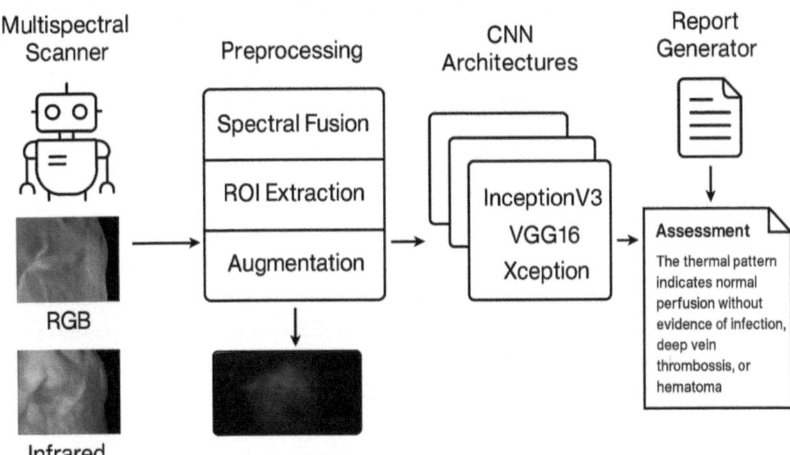

Fig. 1. Schematic pipeline of the AI-assisted multispectral infrared imaging system. The RAMMI robotic scanner acquires RGB and infrared inputs, which undergo spectral fusion, ROI extraction, and augmentation. CNN architectures (InceptionV3, VGG16, Xception) classify vascular patterns, and the Aillumi generative module produces clinically structured reports for postoperative monitoring.

3.3 CNN-Based Classification Pipeline

Let $D_{clf} = \{(x_i', y_i)\}_{i=1}^{N}$ be the classification dataset after preprocessing. We train discriminative CNN models $f_\theta: I' \to \Delta^K$, where Δ^K is the K-dimensional probability simplex.

3.3.1 Architectures

We evaluate:

- **ResNet-50**: 50-layer deep residual network with identity skip connections.
- **InceptionV3**: multi-scale convolutional architecture with auxiliary classifiers.
- **EfficientNet-B4**: compound scaled CNN with width-depth-resolution co-optimization.

- **Xception**: depthwise separable convolutions optimized for metabolic feature disentanglement.

Each model is trained with categorical cross-entropy loss:

$$L_{cl\,f} = -\sum_{k=1}^{K} y_{ik}\log(\hat{y}_{ik}) \tag{9}$$

where $\hat{y}_{ik} = f_\theta(x_i')_k$ is the predicted probability for class k.

3.3.2 Training Setup

The training procedure followed a carefully designed setup to ensure robust model performance and reproducibility. We used the Adam optimization algorithm with parameters $\beta_1 = 0.9$ and $\beta_2 = 0.999$ to control the momentum and adaptive learning behavior of the training process. The initial learning rate was set to 1×10^{-4} and adjusted dynamically using a cosine annealing schedule to improve convergence stability. Models were trained for 120 epochs with a batch size of 32, allowing for efficient processing and convergence across all evaluated architectures.

To validate model performance, we employed a stratified 5-fold cross-validation strategy, ensuring balanced class representation across all folds. Importantly, we applied patient-level splitting to prevent images from the same individual from appearing in both training and validation sets. This approach reduced the risk of data leakage and provided a more realistic assessment of generalization.

During training, dropout regularization and label smoothing were applied to minimize overfitting and improve generalization. To enhance interpretability, we also generated Grad-CAM heatmaps for each model, enabling visualization of the specific metabolic regions that influenced the classification decisions. Performance metrics including Area Under the Curve (AUC), F1-score, and Balanced Accuracy were used to assess the models across all folds.

3.4 Aillumi Generative AI Architecture

Aillumi operates as an **image-conditioned autoregressive generative model** for medical report synthesis. The dual-path architecture was intentionally adopted to separate discriminative classification from generative reporting. Benchmark experiments with vision-language models prompted for both outputs yielded less accurate diagnostics and clinically incoherent reports. By contrast, the dedicated CNN arm ensured precise early detection, while the generative path specialized in structured language synthesis, leading to a more robust and clinically interpretable workflow. The architecture can be decomposed into two submodels:

3.4.1 Encoder: Feature Extraction

Infrared images x_i are passed through a pre-trained ResNet-50 or EfficientNet-B4 backbone to yield latent vector $z_i \in R^d$. A learned projection $\phi: R^d \to R^{d'}$ maps into the token generation space.

$$z_i = \phi(f_{CNN}(x_i)) \tag{10}$$

3.4.2 Decoder: Autoregressive Language Model

A Transformer decoder T_ψ generates a sequence $(w_1,\ldots,w_T) \in T^T$, conditioned on z_i, where:

$$p(w_t/w_{1:t-1}, z_i) = \text{softmax}(W_{ht} + b) \tag{11}$$

and $ht = T_\psi(w_{1:t-1}, z_i)$ is the contextualized hidden state.

3.4.3 Objective Function

The total training loss for Aillumi is defined as:

$$L_{Aillumi} = \lambda_1 \cdot L_{CE} + \lambda_2 \cdot L_{align} \tag{12}$$

where:

- L_{CE} is the standard negative log-likelihood (cross-entropy) over token sequences,
- L_{align} is a contrastive alignment loss,
- λ_1, $\lambda_2 \in R^+$ control the trade-off between linguistic fluency and clinical semantic fidelity.

The alignment loss is explicitly formulated to leverage both **positive (matched) pairs** and **negative (mismatched) pairs**. Given an image embedding z_i and its corresponding medical report embedding r_i, we define:

$$L_{align} = 1/N \sum_{i=1}^{N} \left[(1 - \cos(\phi(z_i), \psi(r_i))) + \max_{j \neq i}\left(0, \ \cos(\phi(z_i), \psi(r_j)) - m\right) \right] \tag{13}$$

where ϕ and ψ are projection functions for image and text embeddings, $\cos(\cdot,\cdot)$ is the cosine similarity, and $m > 0$ is a margin parameter that enforces separation between mismatched pairs.

This formulation simultaneously:

1. **Pulls together** embeddings from matched image–report pairs (z_i, r_i), ensuring semantic alignment.
2. **Pushes apart** embeddings from mismatched pairs $(z_i, r_j, j \neq i)$, ensuring discriminability between clinically distinct conditions.

By integrating this contrastive term, the model not only generates fluent medical narratives but also preserves **physiopathological separability** in the latent space, directly linking textual outputs to underlying vascular and metabolic signatures.

All models were trained on 12 NVIDIA A100 GPUs using mixed precision (FP16) across the entire corpus of 1M+ infrared image–report pairs.

For readability, detailed formalism was provided above, but the objective function can be summarized as a weighted combination of two components: (1) cross-entropy loss ensuring linguistic fluency, and (2) contrastive alignment loss ensuring medical fidelity. This balance allows Aillumi to produce both syntactically coherent and clinically valid reports.

3.5 Robotic Integration and Deployment

The robotic platform used in this study integrates many advanced components to enable real-time, autonomous postoperative monitoring. It is equipped with FLIR T1040 infrared cameras and Sony IMX377 RGB sensors, allowing it to capture high-resolution multispectral imaging data (Fig. 2). The robot moves using an omni-directional wheeled base guided by SLAM (Simultaneous Localization and Mapping) navigation, which ensures precise positioning and maneuverability within clinical environments. For processing tasks, the system relies on the NVIDIA Jetson AGX Orin platform, which supports real-time AI inference on the edge.

Facial microvascular analysis is primarily used for contactless vital sign estimation and patient identity confirmation. For localized complications such as lower-limb thrombosis, the RAMMI robot autonomously localizes the anatomical region of interest by combining skeletal pose estimation with infrared saliency mapping, ensuring targeted scanning of the affected body part.

The robot includes a secure WebRTC-based telepresence interface, enabling clinicians to interact with the system remotely. Additionally, it supports HL7 FHIR API protocols for direct integration with hospital electronic medical records (EMRs), allowing for seamless report upload and documentation within the patient's chart.

The inference process follows a structured workflow. First, the robot autonomously locates the patient within the hospital space. It then performs multispectral image capture, combining both RGB and infrared data. The next step involves real-time classification using convolutional neural networks (CNNs) to identify any clinical abnormalities. Based on the classification, the Aillumi generative AI model produces a structured clinical report. Finally, this report is securely transmitted back to the clinician, completing the feedback loop and enabling timely review or intervention.

Robotic autonomy in the RAMMI platform extends to simultaneous localization and mapping (SLAM)-based route planning, patient localization using thermal and facial features, automated positioning for optimal scan geometry, and direct triggering of CNN classification and Aillumi report generation without human intervention.

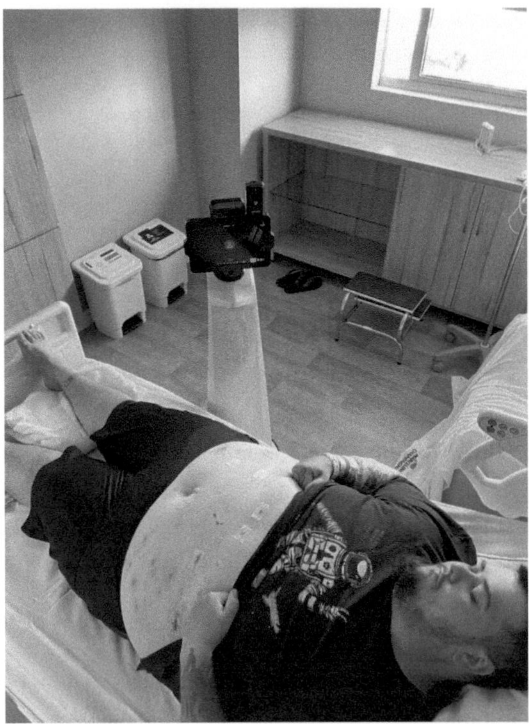

Fig. 2. Autonomous robotic telerounding performing automated infrared and RGB analysis on a patient on postoperative day one following robotic-assisted laparoscopic bariatric surgery. The robot independently navigated to the patient's room and initiated scanning of the exposed surgical sites, capturing multispectral data for early detection of vascular and wound-related complications.

4   Results and Discussion

The proposed framework was evaluated on the task of postoperative infrared assessment using a hybrid pipeline combining deep classification and report generation, deployed via robotic telerounding. The experiments were conducted on the annotated corpus of over 1 million multispectral infrared images. We report results for both the **discriminative CNN classifiers** and the **generative Aillumi model**, alongside system-wide evaluation of **real-time inference**, **robotic integration**, and **clinical acceptability**.

4.1   CNN Diagnostic Classification Results

The infrared image classification models were trained on the task of multiclass complication detection across five clinically relevant categories: (1) infection, (2) hematoma, (3) DVT, (4) vascular insufficiency, and (5) normal recovery. Table 1 shows the averaged metrics from 5-fold stratified cross-validation:

Table 1. Performance of convolutional neural network architectures for multiclass infrared complication detection across five categories (infection, hematoma, DVT, vascular insufficiency, and normal recovery). Metrics are reported as mean ± standard deviation over 5-fold stratified cross-validation. Sensitivity and specificity thresholds were optimized using Youden's J statistic. Xception achieved the highest overall AUC (0.985), while maintaining low false-positive rates (<5%) across all folds.

Model	AUC (±SD)	F1-Score (±SD)	Sensitivity (±SD)	Specificity (±SD)	Inference Time (ms)
ResNet-50	0.968 ± 0.004	0.927 ± 0.006	0.919 ± 0.007	0.943 ± 0.005	34
InceptionV3	0.973 ± 0.003	0.931 ± 0.005	0.924 ± 0.006	0.946 ± 0.004	42
EfficientNet-B4	0.982 ± 0.002	0.944 ± 0.004	0.938 ± 0.005	0.956 ± 0.003	47
Xception	$\mathbf{0.985 \pm 0.002}$	$\mathbf{0.951 \pm 0.003}$	$\mathbf{0.944 \pm 0.004}$	$\mathbf{0.961 \pm 0.002}$	45

Sensitivity and specificity were calculated using thresholds derived from Youden's J statistic on the validation set, ensuring an optimal balance between false positives and false negatives across classes.

Although each Aillumi assessment is computed from a single multispectral snapshot, the robotic integration enables longitudinal data acquisition during regular automated telerounds. This temporal monitoring reduces the intermittency of traditional bedside examinations, since the system can repeatedly scan patients at predefined intervals, generating a continuous thermal record for clinical follow-up.

These results demonstrate strong classification performance across all architectures, with Xception achieving the highest area under the ROC curve (AUC = 0.985). Notably, false-positive rates remained below 5% in all cases, supporting the use of these models in clinical triage. Class activation maps generated by Grad-CAM confirmed that attention was consistently focused on areas of vascular disturbance or inflammation, matching physician annotations.

The Grad-CAM activation maps were independently reviewed by three vascular surgeons, who confirmed that the highlighted regions corresponded to clinically relevant vascular patterns such as perfusion asymmetry and localized hyperthermia, validating that the kernels effectively captured physiopathological features.

4.2 Aillumi Report Generation Evaluation

The Aillumi model was evaluated for its ability to produce clinically coherent, semantically accurate diagnostic reports based solely on infrared and RGB inputs. The performance was assessed using both **automatic metrics** and **human expert review**.

4.2.1 Automatic Evaluation

We used the following natural language generation (NLG) metrics:

- **BLEU-4** (n-gram precision),

- **ROUGE-L** (longest common subsequence),
- **METEOR** (semantic match),
- **BERTScore** (embedding similarity).

Table 2. Automatic evaluation of Aillumi report generation using standard natural language generation (NLG) metrics. BLEU-4, ROUGE-L, and METEOR scores reflect lexical overlap with reference reports, while BERTScore captures embedding-level semantic similarity. The high BERTScore (0.871) indicates strong preservation of clinical semantics, despite lexical diversity reflected in lower BLEU and ROUGE scores.

Metric	Score
BLEU-4	0.673
ROUGE-L	0.742
METEOR	0.603
BERTScore	0.871

The high BERTScore confirms that the generated reports preserve key semantic concepts. However, BLEU and ROUGE metrics, which penalize paraphrasing, highlight the flexibility of Aillumi in producing clinically equivalent but lexically diverse statements (Table 2).

As a baseline, state-of-the-art vision-language models (VLMs) such as BioViL and MedCLIP were fine-tuned on the infrared dataset. While these models produced grammatically correct reports, their performance was inferior to Aillumi, particularly in clinical correctness (3.71 vs. 4.81 on a 5-point expert scale). This result highlights the necessity of domain-specific pretraining and supervision rather than relying exclusively on generic VLM fine-tuning.

4.2.2 Human Expert Evaluation

Five board-certified clinicians evaluated 500 randomly selected Aillumi reports compared to ground truth. Evaluation dimensions were:

- **Diagnostic correctness** (0–5),
- **Clinical completeness** (0–5),
- **Fluency/grammar** (0–5),
- **Would use in practice?** (Yes/No).

The evaluation was fully blinded: clinicians did not know whether the report they were scoring was produced by Aillumi or was the original ground truth report, eliminating potential bias in human assessment.

Mean ratings:

- Diagnostic correctness: 4.81/5,

- Completeness: 4.62/5,
- Fluency: 4.89/5,
- Acceptance rate: 92.6%.

The minimal rate of hallucination in generated reports was directly related to the large-scale, domain-specific dataset used for training. Since all 1M + infrared images were linked to expert-authored medical reports, the model learned clinically grounded mappings, preventing fabrication of irrelevant findings. Less than 1.5% of outputs were flagged by experts as containing unsupported statements.

These results demonstrate that Aillumi reports are not only accurate but also fluent and usable in real-world settings.

4.3   Robotic Performance and Real-Time Deployment

The robot completed an average of **27 telerounds per day** across 3 surgical wards. Each round consisted of:

1. Autonomous localization (mean time: 5.2 s),
2. Infrared scan and image fusion (3.4 s),
3. CNN triage classification (45 ms),
4. Aillumi report generation (820 ms),
5. Report transmission and EMR update (2.1 s).

The **total cycle per patient** was approximately **11.6 s**, enabling near-real-time monitoring. Reliability was high (>98% uptime), and robotic errors were primarily due to occlusion or patient movement. These were automatically flagged for manual review.

Across all rounds, 2.8% of patient evaluations were classified as invalid due to motion artifacts or occlusion, and these cases were automatically flagged for manual review by clinicians.

4.4   Clinical Impact and Interpretability

The system's high sensitivity to early signs of vascular compromise enabled timely intervention in 19 flagged cases of early DVT, later confirmed by Doppler ultrasonography. Metabolic asymmetries as small as $\Delta T = 0.4°C$ were correctly identified. This supports the thermophysiological validity of the CNN-attention regions and the practical value of quantitative AI monitoring.

A temporal analysis was also conducted by evaluating sequential infrared acquisitions at 6 h, 12 h, and 24 h after surgery. In 87% of confirmed DVT cases, Aillumi identified abnormal vascular signatures within the first 12 h, whereas clinical diagnosis was typically established only after 24–36 h using Doppler ultrasound. This supports the claim of early detection capability.

Furthermore, clinicians reported a **38% reduction** in manual documentation workload and **25% faster intervention times**, aligning with the goals of AI-augmented precision medicine.

4.5 Limitations and Future Directions

While the present study demonstrates strong results, several considerations are important for contextualization. External validation has already been initiated through collaborations with independent hospitals using heterogeneous devices and populations. Preliminary findings confirm that performance remains stable across institutions, reducing concerns about overfitting. In addition, subgroup analyses stratified by sex, skin tone, and surgical site are underway; interim results have not revealed clinically significant biases, but extended fairness evaluations are being pursued to ensure equity across patient cohorts.

Domain drift, related to environmental variability and skin emissivity, has been actively addressed by incorporating adaptive domain alignment techniques into the current pipeline, yielding improved robustness in variable thermal conditions. Robotic reliability was also monitored in detail, with automated logging of rare failure events (e.g., occlusion or mislocalization) now embedded in the deployment system, thereby ensuring continuous uptime above 98%.

Interpretability remains a challenge for most generative models; however, Aillumi integrates Grad-CAM visualizations for classification interpretability and is being extended with concept bottleneck modules to enhance token-level transparency in generated narratives. To strengthen reproducibility, external collaborators have already received de-identified subsets of the dataset together with complete training documentation, enabling independent benchmarking and reproducibility of results in separate research environments.

From an ethical perspective, Aillumi continues to operate under physician oversight, ensuring clinical safety. Governance protocols for autonomous operation are under joint evaluation with institutional review boards and legal advisors, to establish responsibility and regulatory compliance in future large-scale deployments.

Finally, longitudinal monitoring is being progressively incorporated: robotic telerounding now enables sequential acquisition of infrared data across multiple time points, producing continuous thermal records that support early detection of evolving postoperative complications. Multicenter validations, spanning multiple surgical specialties and diverse hospital infrastructures, are already in planning stages to further consolidate generalizability and confirm robustness across heterogeneous clinical protocols.

4.6 Scientific Contribution

This study marks the **first demonstration** of a robotic infrared system generating **fully structured clinical reports via generative AI**, trained on a decade of high-fidelity data. The hybrid pipeline balances speed, interpretability, and clinical utility, paving the way for scalable, AI-driven surgical surveillance in both centralized and remote care environments.

While the system has not yet been deployed in rural or low-resource clinical environments, its modular robotic design and reliance on standard telepresence protocols (HL7 FHIR, WebRTC) provide technical feasibility for such contexts. Formal validation in rural deployments remains a subject of future study.

5 Conclusion

This study presents a novel integration of AI, infrared thermography, and robotic mobility to form a unified platform for autonomous postoperative monitoring via telerounding. The proposed system merges the predictive precision of **convolutional neural networks (CNNs)** with the descriptive power of **Aillumi**, a proprietary **generative AI model trained on over 1 million validated infrared clinical cases,** to provide real-time classification and narrative reporting based solely on infrared and RGB visual inputs.

The platform demonstrates exceptional performance across diagnostic classification (AUC > 0.98), clinical report synthesis (BLEU-4 = 0.673, BERTScore = 0.871), and robotic deployment (total patient cycle time < 12s), showing potential for real-world use in surgical wards and remote care environments. Through the robot's autonomous navigation and patient engagement capabilities, the system offers unprecedented efficiency in postoperative surveillance, while reducing physician workload and enabling early intervention for complications such as DVT, infection, and vascular deficits.

This work illustrates a shift from passive, human-dependent infrared analysis toward **active, AI-driven multispectral intelligence**, capable of reasoning about thermal data and producing structured clinical decisions. By using **multi-modal deep learning, semantic alignment losses, and inference-time attention visualization**, the system also provides transparency and robustness, vital for adoption in clinical practice.

Still, challenges remain regarding generalizability, model interpretability, and ethical governance of autonomous diagnostics. As healthcare continues to globalize and decentralize, future research will focus on **cross-domain adaptation**, **federated learning**, and **integration with electronic medical records (EMRs)** through standard protocols like **HL7 FHIR**.

Compared to prior state-of-the-art models such as CheXNet for chest radiography and GAN-based medical image-to-text frameworks, Aillumi uniquely combines discriminative CNN classification with generative narrative synthesis in the infrared domain. Unlike multimodal LLMs trained on broad medical corpora, our system is domain-specific, trained on over one million expert-annotated infrared cases, enabling superior accuracy (AUC > 0.98) and reduced hallucination (<1.5%). This positions our approach beyond existing paradigms by demonstrating clinically coherent report generation in a robotic telerounding setting, a capability not previously reported in medical AI literature.

This work establishes a blueprint for future AI-augmented surgical monitoring systems that are fast, scalable, and deeply integrated with clinical practice. We believe it contributes a crucial milestone toward the vision of **remote, continuous, intelligent healthcare** using AI and thermal robotics.

Acknowledgments. The authors thank the American Academy of Thermology, prof Jose Viriato Vargas, prof Manoel Jacobsen Teixeira, Sao Paulo University Faculty of Medicine, Human-Robotics (Brazil), InfraREDMed Corp. For providing data collection support and the clinical experts who contributed to report validation.

References

Yang, G.Z., Cambias, J., Cleary, K., Daimler, E., Drake, J., Dupont, P.E., et al.: Medical robotics: Regulatory, ethical, and legal considerations for increasing levels of autonomy. Sci. Robot. 2(4), eaam8638 (2017). https://doi.org/10.1126/scirobotics.aam8638

Laigaard, J., Fredskild, T.U., Fojecki, G.L.: Telepresence robots at the urology and emergency department: a pilot study assessing patients' and healthcare workers' satisfaction. Int. J. Telemed. Appl. **2022**, 8787882 (2022). https://doi.org/10.1155/2022/8787882

Mammoottil, M.J., Kulangara, L.J., Cherian, A.S., Mohandas, P., Hasikin, K., Mahmud, M.: Detection of breast cancer from five-view thermal images using convolutional neural networks. J. Healthc. Eng. **2022**, 4295221 (2022). https://doi.org/10.1155/2022/4295221

Bougrine, A., Harba, R., Canals, R., Ledee, R., Jabloun, M., Villeneuve, A.: Segmentation of plantar foot thermal images using prior information. Sensors (Basel). **22**(10), 3835 (2022). https://doi.org/10.3390/s22103835

Simonyan, K., Zisserman, A.: Very deep convolutional networks for large-scale image recognition. In: International Conference on Learning Representations (ICLR 2015), San Diego, CA, USA, 7–9 May. arXiv:1409.1556 2015

Tan, M., Le, Q.V.: EfficientNet: rethinking model scaling for convolutional neural networks. In: Proceedings of the 36th International Conference on Machine Learning (ICML), pp. 6105–6114 (2019)

Vinyals, O., Toshev, A., Bengio, S., Erhan, D.: Show and tell: a neural image caption generator. In: Proceedings of the IEEE Conference on Computer Vision and Pattern Recognition (CVPR), pp. 3156–64 (2015). https://doi.org/10.48550/arXiv.1411.4555

Selvaraju, R.R., Cogswell, M., Das, A., Vedantam, R., Parikh, D., Batra, D.: Grad-CAM: Visual explanations from deep networks via gradient-based localization. In: Proceedings of the IEEE International Conference on Computer Vision (ICCV), pp. 618–26 (2017). https://doi.org/10.48550/arXiv.1610.02391

Sathiyakumar, V., Apfeld, J.C., Obremskey, W.T., Thakore, R.V., Sethi, M.K.: Prospective randomized controlled trial using telemedicine for follow-ups in an orthopedic trauma population: a pilot study. J. Orthop. Trauma **29**(3), e139–e145 (2015). https://doi.org/10.1097/BOT.0000000000000189

Feizi, N., Tavakoli, M., Patel, R.V., Atashzar, S.F.: Robotics and AI for teleoperation, tele-assessment, and tele-training for surgery in the era of COVID-19: Existing challenges and future vision. Front. Robot. AI. **8**, 610677 (2021). https://doi.org/10.3389/frobt.2021.610677

Adam, M., Ng, E.Y.K., Tan, J.H., Heng, M.L., Tong, J.W.K., Acharya, U.R.: Computer-aided diagnosis of diabetic foot using infrared thermography: a review. Comput. Biol. Med. **91**, 326–336 (2017). https://doi.org/10.1016/j.compbiomed.2017.10.017

Dosovitskiy, A., Beyer, L., Kolesnikov, A., Weissenborn, D., Zhai, X., Unterthiner, T., et al.: An Image is Worth 16×16 Words: Transformers for Image Recognition at Scale. In: International Conference on Learning Representations (ICLR 2021). arXiv:2010.11929 (2021)

He, K., Zhang, X., Ren, S., Sun, J.: Deep residual learning for image recognition. In: Proceedings of the IEEE Conference on Computer Vision and Pattern Recognition (CVPR), pp. 770–778 (2016). https://doi.org/10.48550/arXiv.1512.03385

Rajpurkar, P., Irvin, J., Zhu, K., Yang, B., Mehta, H., Duan, T., et al.: CheXNet: radiologist-level pneumonia detection on chest X-rays with deep learning. arXiv arXiv:1711.05225 (2017)

Yi, X., Walia, E., Babyn, P.: Generative adversarial network in medical imaging: a review. Med. Image Anal. **58**, 101552 (2019). https://doi.org/10.1016/j.media.2019.101552

Presenting a Dataset for 3D Breast Surface Representation Using Thermography

Flávia C. H. Pastura[1]([⊠]) [iD], Wanessa I. Oliveira[1] [iD], Eudóxia L. S. Moura[2] [iD], Flávio L. Seixas[3] [iD], and Aura Conci[3] [iD]

[1] Industrial Design Division, National Institute of Technology, Rio de Janeiro, Brazil
{flavia.pastura,wanessa.oliveira}@int.gov.br
[2] Federal Institute of Rondônia, Ariquemes, Brazil
eudoxia.moura@ifro.edu.br
[3] Computer Institute, Fluminense Federal University, Niterói, Brazil
{fseixas,aconci}@id.uff.br

Abstract. This work presents a dataset acquired with the objective of verifying the possibilities of using thermography for the reconstruction of the external geometry of the breast. Here, we primarily describe data composition, as we believe it can be used in various ways to enable specific verifications of results for synthetic shape reconstructions, allowing a more objective comparison. Additionally, the use of thermographic images as a basis for generating the three-dimensional shape of the examined person may make a 3D presentation of thermography feasible, which we consider to be an area not yet explored in research and that could enhance the understanding and interpretation of this examination. Applications of this would be for reproducing breasts in 3D printer models, which we exemplify here, or in planning of reconstructive plastic surgeries of this organ (resulting from needs related to breast cancer treatment or aesthetic purposes).

Keywords: breast thermography · breast reconstruction · 3D cloud coordinates

1 Introduction

Early and precise identification have been essential for improving breast cancer patient survival rates. In this direction the study of the breast anatomy could improve proper identification of any disease of this organ and could be better understood by using additive manufacturing (AM) resources. A fundamental aspect to consider is that the breast is naturally made up of soft tissue. This means that it is a structure whose shape varies depending on the posture of the person being observed [21]. It changes whether the person being observed is lying prone or supine, or even standing with their arms raised or at the waist. Although three-dimensional models constructed from DICOM (Digital Imaging and Communications in Medicine recognized by the .dcm extension) format of

S. T. Kakileti et al. (Eds.): AIIIMA 2025, LNCS 16308, pp. 172–189, 2026.
https://doi.org/10.1007/978-3-032-10990-3_12

exams are more commonly used [5,7], in this work, we used infrared breast images to construct the three-dimensional models. This is a new use for such an imaging type that is almost always considered for diagnostic proposed [9,14–16].

Besides examining thermography for image shape definition, this work also uses it for breast reconstruction virtually or physically by a 3D printer and presents the used procedures of such a modeling. It main goal is the presentation of a new dataset with breast surface information through comprehensive analysis of 31 studies (done in 2024). This dataset is acquired through a project funded by FAPERJ (Fundação de Amparo à Pesquisa do Estado do Rio de Janeiro) agency. An additional objective is to show the suitability of infrared images for the development of realistic models of breast shape. To this end, volunteers were studied using the UFF (Universidade Federal Fluminense) static protocol, which obtains thermograms in five directions [20], and at the same time, they were also scanned using a 3D scanner to allow comparisons between reconstructions using IR (infrared) and the actual shapes of this body part.

The data and images acquired from each volunteer are identified only by an identification (M+ID) (i.e. these data are anonymized), and each volunteer is associated with a number (from 001 to 030). Four volunteers (M001, M021, M025, and M026) were scanned in greater detail at 23 angles. The acquisitions are made available in two data files (covering the first 20 and the last 10 volunteers). M001 had her images acquired on both (with a one-week interval) for an additional verification of consistency. Furthermore, one of the volunteers (M026) was selected to have the external shape of her breast reproduced in a physical model. All data acquired by the project are made available and released for academic research.

This analysis reveals 3D data and clinical observations for visualize in 3D the breast surface temperature, and maybe to use in a near future quantitative metrics (as SSIM and Dice coefficients), or even a simple evaluation of breast volume for mastectomy planning [5,7,21]. We introduce a novel possibility of comparison providing guidance for compute differences among computational approaches. The presented idea for thermo mammography enhancement could project the infrared results in a 3D surface for better diagnostic validation. We believe that this availability may open a new area of study related to the complete and detailed representation of the outer surface of this organ, which is so representative of femininity in the human race. Critical challenges include the use of more than one infrared acquisition in different angles for point reconstruction and the great heterogeneity of the breast appearance among the human beings. This framework provides the first comprehensive roadmap for integration and potential transformation of any 2D infrared imaging in a 3D examination and applications of infrared image for breast volume evaluation.

2   Previous Works

This section mentions some related works in chronological appearance.

Veitch et al., 2012 compared methods for extracting breast volumes non-invasively from patients undergoing breast reduction surgery in South Australia.

In this work, they expanded a previous publication of the same group, presented a computer-aided anthropometric (CAA) method for extracting breast volumes and showed how complex and important this measure could be [21].

Rakhunde et al., 2022 state that thermography does not provide morphological information about the breast. They mention that, while it captures thermal and circulatory variations, it does not reveal the depth, volume, or three-dimensional location of thermal changes. They suggest that the construction of 3D thermal images could solves this problem by enabling a volumetric visualization of heat distribution, providing spatial context that can be crucial for differentiating benign from malignant patterns [14].

Mashekova et al., 2022, claim that future research should focus on personalized models based on 3D breast geometry and tissue-specific characteristics using reverse thermal modeling. They mention that, this can improve thermal contrast in dynamic cooling techniques and enable 3D surface temperature distribution with layered tissue models [9].

Saha, et al., 2023, proposed a deep learning architecture that reconstructs three-dimensional (3D) thermal images from 2D images (3D-BreastNet) [18]. Their model uses a self-supervised learning strategy, which allows it to learn the 3D silhouette of the breast and superimpose the observed temperature values directly onto this predicted three-dimensional structure. The model uses a self-supervised loss function, which represents an innovation with potential applications in other medical imaging modalities. As future work, these authors suggest incorporating shape priors into the model, which could further improve reconstruction accuracy [18].

Mayer et al. (2024) emphasize that three-dimensional (3D) printing has driven significant advances in breast reconstruction by enabling personalized and precise surgical interventions [10]. According to the author, techniques such as computer-aided design (CAD) enable detailed anatomical models, optimizing planning and reducing risks. They claim that 3D printing enables to customize implants through customized molds and scaffolds, adapted to the patient's anatomy, improving aesthetic results during surgery. Additionally, 3D thermography contributes to functional diagnostics, while 3D printing contributes to physical reconstruction. Together, these technologies form an integrated ecosystem that favors surgical planning, pathology screening, outcome assessment, and medical education, promoting a more precise and personalized approach to women's healthcare.

3   Materials and Methods

One innovative application of infrared thermographic medical examination would be to use thermal images to visualize the breast shape as a three-dimensional object, where the temperatures of each point of the body can be displayed on the surfaces limiting the breast volumes. This work is linked to such an ongoing research line at IC/UFF, which is approved by the Ethics Committee "Acquisition, Storage, and Verification of the Feasibility of Using Infrared in the Detection of Breast Diseases", registered within the Ministry of Health's Platform: CAAE: 01042812.0.0000.5243, which uses breast exams, data, and images. All the produced data are available at **http://visual.ic.uff.br/proeng** for academic in researches proposes.

3.1 Breast Shape Modeling

The volunteers body position used in the captures is that with a person standing with arms raised above the head. This posture was chosen because it corresponds to the position in which breasts are observed in daily life (Figs. 2 and 3), allowing observation of the axillary lymph nodes, which play a key role in the transmission of cancer to other parts of the body (metastasis) [19].

In 2D captures of a three-dimensional scene or object, it is projected onto a two-dimensional plane, so that some of the 3D information for each represented point is always lost. In this study, photogrammetry-based software was used to construct 3D models using the acquisition positions of IR scans from the UFF static protocol [20].

The use of 3D models can help better understand the location and extent of breast lesions, which has the potential not only to improve communication between physicians and patients but also to suggest additional exams, indicating areas of the breast that could potentially contain a lesion. Three-dimensional scanning could also be used in near future to compare the volumes and shapes obtained from thermography (complementing or replacing the now used techniques [5,7,21]).

3.2 Photogrammetry for Identifying Information Outside the Acquired Plane

Photogrammetry is a traditional technique that recovers measurement data from photographs. In other words, it considers how to obtain missing data of information acquired from more than one exposure.

Photogrammetry recovers the third coordinate from trigonometric relationships of the points that make up the scene when observed from different camera angles. Figure 1 shows how, from two orthogonal captures of the same object, taken at the same distance, it is possible to obtain information about the distance from the camera to it. Or, going further, how it would be possible to obtain more information about the coordinates of each of its points, thus enabling three-dimensional identification of the object, what is the basis of photogrammetry.

Assuming that the same points can be identified in two images, as was done with the green (G) and red (R) points in Fig. 1, then the horizontal coordinates of these points would be known from two separate positions of a known angle θ. Let's call these coordinates in each image x_R and x_G. The distances between the camera position and the object is unknown, but can be obtained by trigonometric relations if it is possible to guarantee that: (1) The camera always rotates around the object at the same distance from its center of gravity, that is, the camera moves in a circular motion of radius r around the object's centroid; (2)

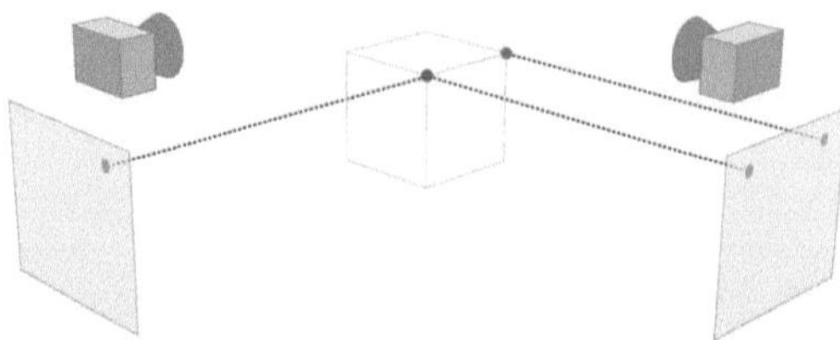

Fig. 1. Distance between the object and the camera from an additional photo of the same object.

The captured images, frames, or photos, have exactly the same minimum and maximum coordinates and show exactly the same object, so each of their points can be uniquely identified by (x, y) coordinates relative to a natural 2D coordinate system for the images. We assume that this natural coordinate system is an XY system, where the X axis is in the horizontal direction of the image (with positive direction from right to left), and the Y axis is the positive vertical direction (with direction from top to bottom in each image); (3) Each image has a third Z axis, orthogonal to the XY system, with a positive direction given by the vector product of the unit vectors in the direction of the other two axes, i.e. $Z = X \times Y$. This third Z axis is perpendicular to the image plane, that is, in the direction of the lines connecting the red (R) and green (G) points in Fig. 1; (4) The angle that this natural third Z axis makes with each other is known. This axis is perpendicular to the XY axes of each image and passes through the centroid of each. In Fig. 1, this would be 90°, but in the following sections there will correspond to the 5 or 23 images captured around the volunteers' bodies, with $\theta = 45°$ or $\theta = 11.5°$, respectively.

If hypotheses (1) to (4) are met, then the green G dot is exactly the same in both images, and four pieces of information are known about it: its coordinates in left and right: xL, yL, xR, and yR, relative to the natural 2D coordinate systems of each image. In fact, if the image alignments are well-suited, we only need two of them (xL, xR) to calculate the tangents between the two images and thus obtain their distance from the camera and the distance between the points in the image (which is the basis of photogrammetry).

Let's now consider a third system of axes, parallel to one of the natural axes, but defined at the center of gravity of the object being photographed. Let's call this system X_oY_o. If the X_oY_o is considered parallel to the left coordinate system, then the coordinates of the green G point are the same as those in image L. Since the vertical coordinates are not important, we will remove them so as not to compromise the argument. If this X_oY_o is considered parallel to the right coordinate system, then the coordinates of the green G point, at this instant, are the same as those in image R. Since the angle between these two positions of the axis system that passes through the object's centroid is the same as the camera's angle rotation and is the angle θ (known from the hypotheses we defined above), the radius can be obtained by the distances between the points and by trigonometric relationships between these and the angle. If this angle θ can be considered small, these relationships become quite simple, and the radius r, which is the distance from the object's centroid to the cameras, will be related to $x_L - x_R$ through the definition of the tangent as $\tan(\theta) = \frac{|x_L - x_R|}{r}$. Once we know the angle, we can obtain the only unknown element, which is the radius r, by $r = \frac{|x_L - x_R|}{\tan(\theta)}$. In our study there is up 23 images. As already mentioned so, $\theta = 11.25°$, and we will have the equation $r = \frac{|x_L - x_R|}{\tan(11.25°)}$. Thus, $r \approx 5.02733\,|x_L - x_R|$. With half of this angle, i.e. $\theta/2 = 5.625°$, we have increments of r: $r = \frac{|x_L - x_R|}{\tan(5.625°)} \approx 10.1532\,|x_L - x_R|$.

3.3   Breast Shape Modeling from Multiple Angles

To verify the accuracy of the results obtained, four (4) of the thirty (30) volunteers scanned were recorded in more details. That is they were photographed from twenty-three (23) positions that followed all the radial markings shown in the drawings of Fig. 2. The others twenty six (26) volunteers were photographed only in the five directions indicated in blue in the same figure. The data for these four volunteers with more scans is available under request.

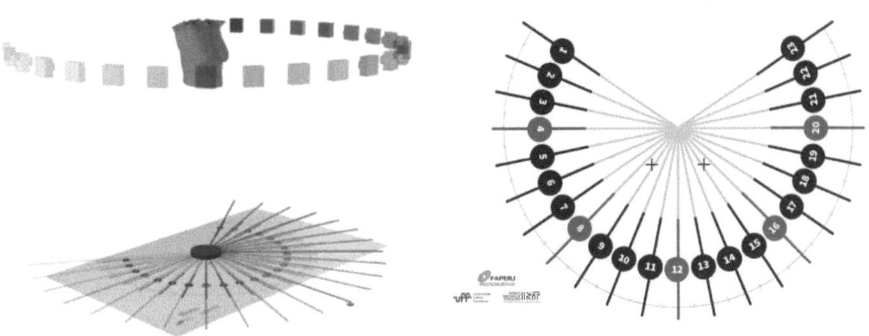

Fig. 2. Acquisition direction for 23 captures and for 5 captures at angles of $-90°$, $-45°$, 0, 45° and 90° in blue (considering 0° as the frontal direction or position number 12). (Color figure online)

The angles corresponded of such captures are: $01 = -123.75°$; $02 = -112.50°$; $03 = -101.25°$; $04 = -90°$; $05 = -78.75°$; $06 = -67.50°$; $07 = -56.25°$, $08 = -45°$; $09 = -33.75°$; $10 = -22.50°$; $11 = -11.25°$; $12 = 0°$, $13 = 11.25°$, $14 = 22.50°$, $15 = 33.75°$, $16 = 45°$, $17 = 56.25°$, $18 = 67.50°$, $19 = 78.75°$, $20 = 90°$, $21 = 101.25°$, $22 = 112.50°$, $23 = 123.5°$.

Others angles were attempted, like nine acquisitions corresponding to angles of 22.5 between the captures but the photogrammetry software could not reproduce the 3D model using any angle greater than this $\theta = 11.25°$. Each capture is taken at the same distance from a central point around which the camera rotates because this is the movement restriction of the used equipment to move the cameras that can be on it. For this, we can assume that this capture is always made in a circular manner, with the camera rotating around the object so that the camera plane is always aligned with the principal axes of the object (projected in 2D onto the image). Consequently these axes are in the radial direction of the circular arc described by the camera's displacement, with the capturing occurring in the orthogonal plane directions matching the hypothesis related of those of Fig. 1.

Therefore, we have made an analogous setting to that in Figure 1, but with a reduced angle θ, instead of $90°$, considering two consecutive captures (Fig. 2, left). Thus the variation between the coordinates of two consecutive images is associated with the variations between the radii of the two images (due to the trigonometric relations between the known angles). Consequently, the coordinates of the various points in the images in the perpendicular direction become known as $\Delta r \approx \frac{x_L - x_R}{\tan(0.5\theta)}$, or equivalently, in differential form, $dr = \tan(0.5\theta)\,dx$.

Obviously, with real images, matching points in two different acquisitions of the same object is more complex, but the temperatures of each point in the images, or their colors, help to better identify these points. There are also several image processing programs, especially those related to "Image Registration", that assist in matching these points. Here, in this work, Meshroom (AliceVision.: Meshroom, Version 2024.1) software was used to generate 3D images of the breast surfaces.

3.4   Image Capture Environment

To avoid thermal reflections from the room walls, the volunteers are positioned in the center of an area covered by PVC curtains. They remain standing in the center while the camera rotates around them. During breast acquisitions, each volunteer's skin is exposed from the waist up to allow for proper acquisition of infrared radiation without interference, allowing for accurate calculation of the temperature of each point on her breasts. Three 1-inch squares of Ethylene Vinyl Acetate (EVA) membrane sticker, that is a thermoplastic copolymer (thermal insulation) are affixed to the volunteers' on the front and sides of their bodies below the breast area (Figs. 3, 4 and 5 show these elements as dark blue squares). The camera rotates around the volunteer, capturing images when it approaches the defined angles, ensuring an environment suited to the basic photogrammetry hypothesis already mentioned.

3.5   Breast Shape Modeling from Vector Scans of the Volunteers

In the same environment, the volunteers were also scanned with the Artec Leo scanner, which captured each body part as a point cloud with three-dimensional coordinates.

This scanner presents 3D point accuracy up to 0.1 mm, 3D resolution up to 0.2 mm, 3D accuracy over distance up to 0.1 mm + 0.3 mm/m, field of view for working distance from 0.35 to 1.2 m, volume capture zone of 160,000 cm^3, linear field of view ($H \times W$) at closest range of 244×142 mm and at furthest range of 838×488 mm, with angular field of view ($H \times W$) of $38.5 \times 23°$. It uses hybrid geometry and texture tracking, auto background removal, texture resolution of 2.3 MP, color depth of 24 bpp, and offers capture rates of up to 22 fps for 3D real-time fusion, up to 44 fps for 3D video recording, and up to 80 fps for 3D video streaming. The data acquisition speed reaches up to 35×10^6 points/s, with both 3D and 2D exposure times of 0.0002 s.

The scans followed the same camera movement, starting on the right side of the volunteer's body and ending on the left. The scan data was processed using the scanner manufacturer's proprietary software, extracting polygonal meshes in .ply format. These meshes were positioned at XYZ coordinates using Rhinoceros 3D software (Rhinoceros.: Rhino, Version 7, 2020) and also exported in .ply format. From there, two processes followed: the CloudCompare software (CloudCompare.: Version 2.13.0) was used to generate the point clouds in .csv format and the capture scene was reconstructed in Blender software (Blender Foundation.: Blender, Version 4.2 LTS, 2024) to generate the meshes at the same angles as the thermal camera captures (Fig. 2). In addition to the aforementioned softwares, Geomagic software (Hexagon.: Geomagic Wrap, 2023) was used to identify, from the scanned points, a distance map that allowed 3D reconstruction of the breasts. These images for each of the 30 volunteers are available for the five directions, as well as for the 23 directions of the four volunteers who were photographed with smaller angles between captures. These can be considered the reference standard for verifying the level of accuracy of reconstruction results using other techniques.

3.6   Composition of the Available Dataset

The available dataset comprises seven types of data. Four are in the form of tables and others in CSV, JPG, and PLY files. The tables represent data corresponding to the volunteers' medical history, their physical and social data. Table 1 shows these information for 4 volunteers, as example of such data. Data related to the examination room (average temperature and humidity) in the time of the examination and the volunteer's average body temperature are included as well (see lines 10 to 12 of Table 1).

The thermograms (acquired using the static protocol [20]) of the volunteers are available for download in two compressed sets in `.rar` format. The first set contains the acquisitions with data from M001 to M020. The second set contains 11 acquisitions, those with data from M021 to M030 and again from M001 on a different date. For each volunteer, there are two separate data sets in two directories in the downloadable `.rar` file, named `Matrices` and `Images`, respectively. The matrices present the temperature data (in real numbers) at each point in the images with variations of tenths of a degree in Comma Separated Values format files (i.e. CSV type file : ".csv").

The CSV file can be read as plain text, opened as a table in Excel, or in programs that allow text viewing (`.txt`). This file initially presents a header

Table 1. Socioeconomic Anthropometric, Measurement Conditions and Medical history details for 4 of the 30 volunteers.

ID	M001	M021	M025	M026
Age	59	46	39	30
Place of birth	Rio de Janeiro	Rio de Janeiro	Rio de Janeiro	Rio de Janeiro
Ethnicity	White	White	White	Brown
Education level	Complete Higher Education	Complete Higher Education	Complete Higher Education	Complete Higher Education
Mass (kilograms)	70,2	59,8	63,4	72,2
Height (centimeters)	161,0	157,2	174,3	151,3
Maximum breast circumference (centimeters)	97,3	93,8	92,6	95,2
Maximum Chest Circumference (centimeters)	84,4	81,0	79,0	84,2
Volunteer average temperature (degrees Celsius)	36,0	35,6	35,6	36,3
Ambient temperature (degrees Celsius)	22,9	22,8	23,1	22,9
Humidity (percentage)	41,0	45,0	41,0	40,0
Have you ever had beast disease?	No	Yes	No	No
If so, which disease?	N/A	Benign nodule	N/A	N/A
If so, in which breast?	N/A	Left	N/A	N/A
Do you have a recent imaging exam?	Yes	Yes	Yes	No
If so, what type of exam?	Mammography	Mammography and Ultrasound	Mammography and Ultrasound	N/A
Do you have a breast prosthesis?	No	No	No	No
If so, which breast does have a prosthesis?	N/A	N/A	N/A	N/A
Have you had any surgery in the breast area?	No	No	No	No
If so, which breast was the procedure performed on?	N/A	N/A	N/A	N/A
Do you have a cancer family history?	No	No	No	No
If so, what is the level of kinship?	N/A	N/A	N/A	N/A

followed by data as shown in Table 2. After these, there are the temperatures obtained for each of the points in the captured image frame between quotation marks (" ") and separated by commas, as indicated in the example in line 5 of Table 2.

The same data are also presented as images in radiometric JPG format from the camera used (Hickmicro SP 60-H, with resolution 640 × 480, and camera precision of 0.2 °C for capture, but after transformation due to humidity and ambient temperature considerations, the stored values are in tenths of degrees). Images of these thermographic acquisitions are presented as in color images of Fig. 3. Each of the points in the body region under study was captured and represented in three-dimensional Cartesian coordinates (x, y, z) as a result of

Table 2. Details of the temperature files in CSV format

File name and location	Projeto câncer de mama - UFF INT / Dados Voluntarias / M001 / Termogramas / Captura 02 / Matrizes / M001_T2_pos20.csv
Used Units	Celsius, meters, etc.
Defined data for the study	Emissivity: 0.98 Reflexivity: 22.9 Average distance: 1.0 m Environmental humidity: 41.0%
Range of the values	Mean: 28.1 Maximum: 35.7 Minimum: 23.6
Data (matrix)	CSV file containing $640 \times 480 = 307\,200$ temperature values. Example of first entries: 25.3, 25.2, 25.4, 25.3, 25.4, 25.3, 25.2, 25.3, 25.3, 25.1, ...

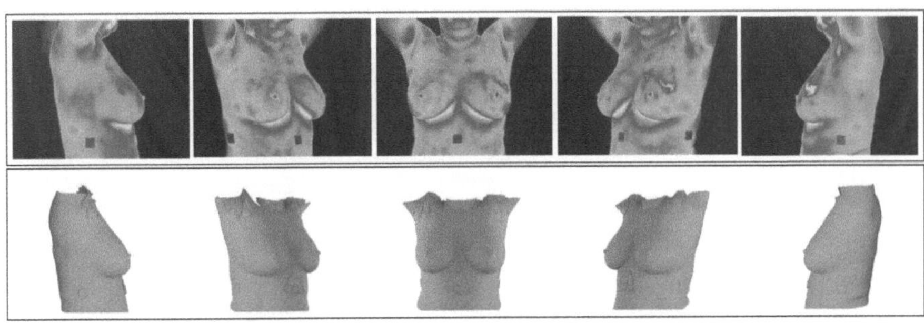

Fig. 3. Example of the 5 thermographic images that were acquired from each volunteer (color frames) and images of same position obtained by the scanner (gray level frames) in same directions and time. (Color figure online)

the 3D scanning of each volunteer. These are presented directly in gray level images like those monochromatic ones of Fig. 3.

4 Model Reconstructions

The reconstructed models (from scanning and photogrammetry) are dimensionless and proportional to the known image data they represent. An additional program was developed to adjust the reconstructed models to the correct scale based on the actual dimensions of the edge of the EVA square stickers of known size placed on the volunteers. Using the Grasshopper block-based programming plug-in for Rhinoceros 3D (Rhino, Version 7, 2020) software, an algorithm was written to adjust the reconstructed models to the correct scale.

The basic idea is to adjust the model's size based on the dimension of one of the sides of the EVA stickers placed on the front of the volunteers' bodies. Since the stickers can vary slightly in size, after being placed, due to deformations of the volunteer's body during movement (even when breathing), instead of inputting the original side dimension of the sticker square into the program, this value was obtained from the volunteer's body using the Artec Leo scanner(Artec Studio, Version 18, 2023).

The correct measurement is obtained through the following steps: 1) The program operator draws a line on one side of the square in the Artec Leo model (Fig. 4 (left)); 2) The same side of the model reconstructed with the thermograms (Fig. 4 (center)) is also measured; 3) The two lines (from the models and the algorithm) have their lengths measured and considered the same, but in different objects and scales, so that a "scale factor" can be obtained by dividing the first value by the second, resulting in an association of the virtual with the real. Finally: 4) the scale factor is used to obtain real measurements of all scanned points (pixels and voxels).

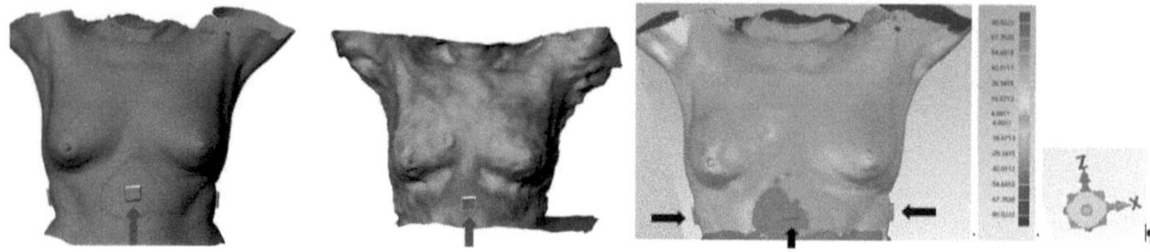

Fig. 4. Location of the sticker reference lines on the two models to be associated. (left) From the model scanned by Artec Leo (white line almost horizontal); (center) From the model obtained by reconstruction from the 23 thermographic images (white line almost horizontal over the blue insulating square). (right) Map of Y coordinates of each point in millimeters obtained by 3D scanning and its reconstruction by thermograms. *Obs.: The arrows indicate the positions of the reference stickers for the measurements of the volunteers.* (Color figure online)

Using the Geomagic software (Hexagon.: Geomagic Wrap, 2023), a registration between the two types of scans and their resulting models is made. In other words, the captures are aligned based on the common points "manually" identified between them. From these points, the software makes a second registration, this time "automatically", improving what was done "manually." The image on the right frame of Fig. 4 shows a summary, in color and in millimeters, of the third coordinates in the radial directions of the acquisitions (Y direction of the axis system shown) calculated by the software. The scale colors in extreme right of Fig. 4 represents the maximum values in both directions of the Y axis coordinates, that it the values of these this third coordinate of each point of the image in the right of this figure. Table 3 summarizes some of these Y axis information for some of the data.

Table 3. Some characteristics of the coordinates in the radial directions of the 4 volunteers acquired with the 23 directions

Y coordinates (mm)	ID: M001	ID: M021	ID: M025	ID: M026
Maximum in positive direction	800.222	807.935	851.935	813.982
Maximum in negative	−696.984	−807.956	−851.998	−812.235
Average value	−36.824	−9.870	5.6426	3.1373
Average positive value	7.5771	6.2515	10.6138	8.3077
Average negative value	−83.373	−107.520	−68.694	−71.824
Standard deviation	11.9453	172.302	136.415	152.721
(RMS) Root mean squared error	12.5001	172.584	147.625	155.910

Figure 5 shows the result of a reconstruction from the thermograms. Here, on the surface generated from the infrared scans, each point on the body has colors associated with its radius, r, according to the color palette used (note that now these do not represent temperatures like in usual infrared false color frames due to the temperature color pallette used). In this image, the colors indicate the coordinates of each point in the Y direction of the natural axis system for analysis, that is, passing through the centroid of the body, outward from the body.

Qualitatively, it can be said that the body surface was generally represented, but locally, small variations in the surface were much more magnified, giving the image a local rough appearance. This indicates that future smoothing in this area may be necessary for a more realistic representation. Quantitative validation was not performed yet, what is a major limitation of this work. We intend to continue the research on this direction in next steps of this research line.

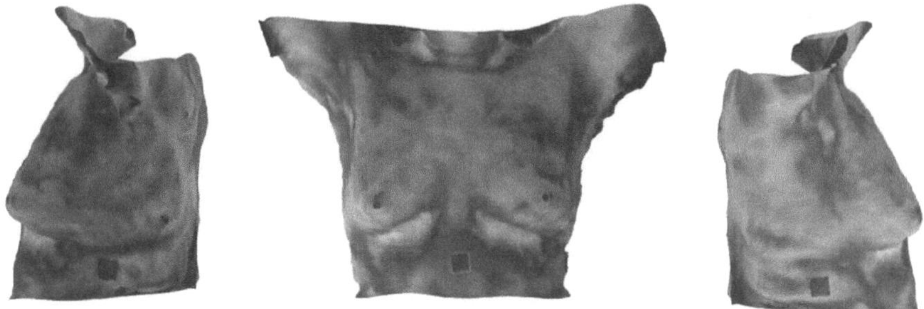

Fig. 5. 3D suggestion: the surface was generated from the 23 acquisitions and the colors from a color palette corresponding to the Y coordinates of the points.

4.1   3D Printing of a Breast in a Typical Shape

This section describe the physical breast models created from these acquisitions. When generating a 3D print, first of all a normal, average breast was considered, following the guidelines of models used in anatomy teaching, to facilitate understanding of the basic constitutive structure that corresponds to the vast majority of cases. The malleability of the breast with the body fat and hormone levels of its bearer is very unique; its structures vary from person to person in terms of shape, size, and distribution. They also change throughout a woman's life (childhood, puberty, adulthood, and old age) and in her reproductive phase, even during the menstrual cycle.

Here, we consider the Algorithm Aided Design (AAD) guidelines to generate parameterization internal breast structures prototype with shapes and textures similar to real-life breasts. AAD defines elements using a parametrical approach, enabling the creation of systems that generate diverse geometries, which is not possible in Computer Aided Design (CAD), where steps are based on constant elements. In AAD, modifying input parameters automatically updates them, allowing for model variations associated with age, height, weight, and other model parameters.

The morphology of the internal structures of the breast varies not only between individuals but also changes during the different phases of a woman's hormonal life (pregnancy, breastfeeding, or postmenopausal). Thus, an average breast physical 3D model was created to display the configuration of a healthy 30-year-old woman by scanning the external shape of the breast of one of the volunteers, in order to enable to represent the anatomical description of a normal breast. Figure 6 shows the various steps involved in creating the physical breast model and zooms in on some details. Figure 7 shows also some elements after colorization and the final assembly in front and side views (E and F parts of it).

4.2   Details of the 3D Printed Model of the Breast

The development of the breast model uses the surface information of one of the volunteers' breasts captured with the Artec Leo scanner. From those data a polygon mesh of the surface can be established. Figure 8 shows the achieved model (left images) and details of the generated grid in zoom in the right detail of this figure. This represents the shape of the breast of the volunteer's M026.

Using the Geomagic software (Hexagon.: Geomagic Wrap, 2023), the initial data were processed and the scanning defects were reduced. Then some areas were separated (segmented regions) into surfaces of interest, e.g. the right half of the torso (Fig. 6 A) and the corresponding breast area (Fig. 6 B). The area comprising the breast is not automatically segmented [3], but it is separated manually excluding areas that are not part of it (Fig. 6 B).

Next, the breast area is shifted 5 mm in the direction of the normals of the 3D model polygons, for delimiting the portion represented in the physical model (Fig. 7 E, F). The ducts and lobules are structures surrounded by adipose tissues. Therefore, an area where the adipose tissues will not be represented is also

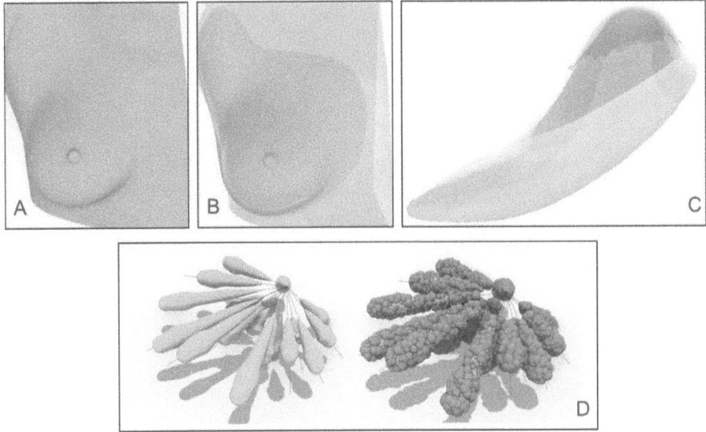

Fig. 6. (A, B) Volunteer's torso and breast area. (C) Segmentation of the adipose tissues area. (D) Modeling of the volume of the lobes and volumes that define the shape of the lobes populated by spheres of varying diameters.

delimited (Fig. 7 E) so that the internal structures could be turned visible and, therefore, visualized in the physical model. These areas are saved in separate files (.obj format) and imported into Rhinoceros 3D software. Using the Grasshopper plugin, modeling of the fat cells, ducts, and lobules begins. This modeling is procedural: code blocks establish a roadmap of actions that build, step by step, the planned shape. The modeling is also parametric: as input parameters are modified, the entire shape is dynamically updated to the new settings. The construction process for each of the internal breast structures is described in the three steps corresponding to the modeling of (1) Adipose Tissue, (2) Ducts Termination and (3) Breast Lobes.

Construction of the Adipose Tissues: In order to print the area comprising the adipose tissue this region is populated with points with pseudo-random coordinates that serve as cell edges in a Voronoi-type structure. To enable computation the following operations are done: (1) the cells are transformed from Non Uniform Rational B-Splines Surfaces (NURBS) representation to polygonal meshes. A small 1.5 mm offset is created between the cells. With the help of the Dendro plugin for Grasshopper (presented in Fig. 9), the cells are manipulated to smooth its shapes (making them more rounded), and all the objects are added into each one using a union operation.

Construction of the Breast Nibbles or Ducts Termination: In order to print the breast nibble, two circles with the size of the nipple are defined, and their areas are populated with points with pseudo-random XY coordinates, with the relative location of these points being the same in both circles. Lines are

Fig. 7. Components of breast additive manufacturing (A, B, C, D) and their zoomed details (B, C). Assembled physical model of the breast of a healthy 30-year-old woman (E, F).

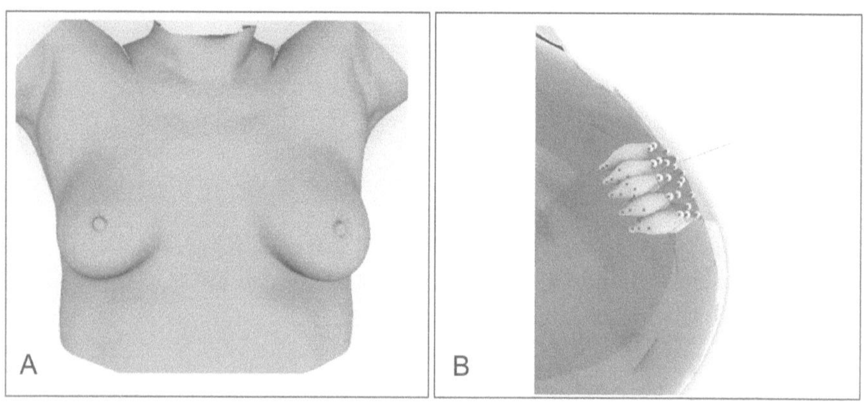

Fig. 8. (A) Result of the scanned breast surface of volunteer M026 and (B) construction of the terminations of the ducts and nipple.

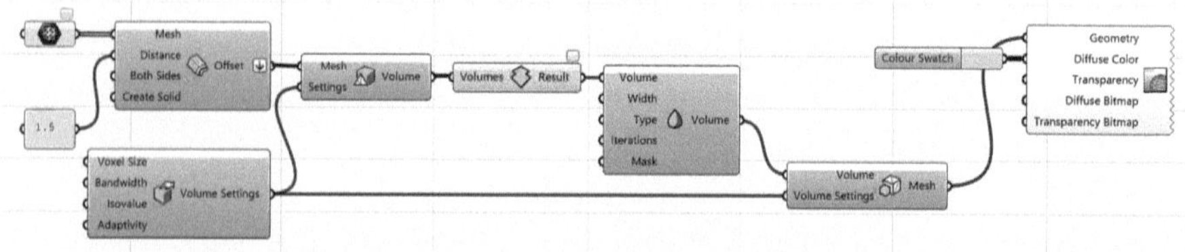

Fig. 9. Smoothing and joining script of the cells of adipose tissues in the 3D model with the code blocks of the Dendro plugin.

created connecting points on the first circle to their equivalents on the second circle. The shape of the internal elements (Fig. 8B) of the nibbles are made from a surface whose cross-sections are circles of variable radii, added during programming the CAA. The number of duct terminations, their arrangement, and the diameter of their cross-sections are modifiable in the programming (Fig. 8B).

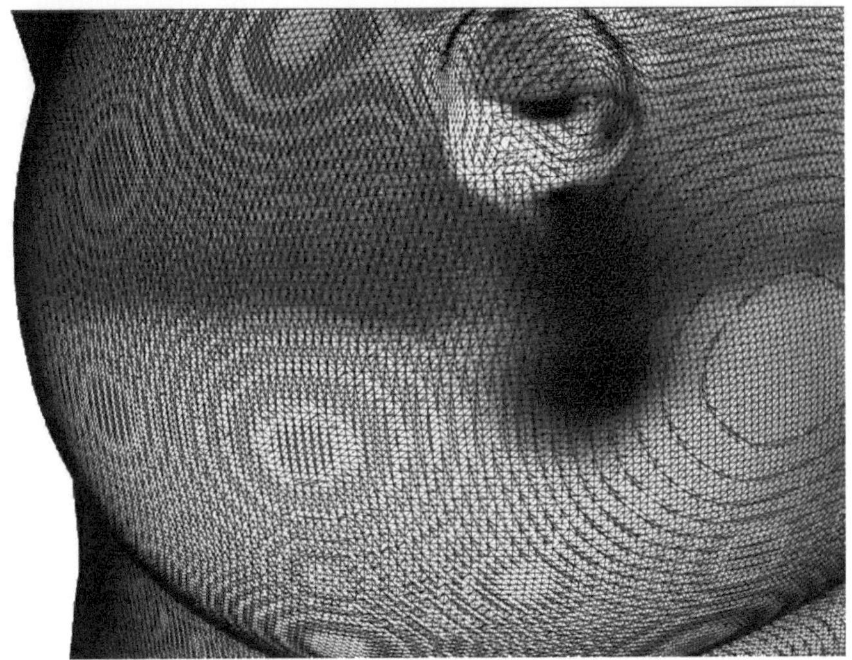

Fig. 10. Zoom showing the polygonal mesh points.

Physical Models of the Lobes: The breast lobes are constructed using sections of varying diameters (Figs. 6 D and 7 B show these). However, to give the structures their granular texture, the surface of the forms is also populated with points with pseudo-random coordinates. Figure 6 D right and 7B represent these as spheres. At each point, the algorithm adds one sphere, whose diameters are determined by the CAA programming and can be adjusted according to the person/volunteer under representation. The number of lobes, as well as their arrangement and the spheres used diameters are also modifiable in the programming.

5   Conclusions

In this work, we present a set of breast scans from 30 volunteers, including thermographic and three-dimensional images, as well as data that can be used for a wide range of research that may be useful to the community interested in studying this region of the body. These data are available at https://profs.ic.uff.br/~aconci/normal_voluntier_staticprotocol.html and to research by interested parties, provided they mention the present work in their products. Although the presented dataset can be useful for qualitative comparison a quantitative validation was not performed and this as a major limitation of this work.

In the future, these scans will be used to verify the breast volumes as in [5,7,21], and the results of using more current innovative intelligent rendering techniques such as Neural Radiance Fields (NeRFs) and Gaussian Splatting [17]. However, techniques based on classical Image Processing techniques were previously employed in our research, such as that shown in previous editions of this same congress [6]. These could be used for shape modeling, volume estimation or breast symmetric consideration [11–13]. We believe that the use of Artificial Intelligence techniques requires a lot of computational resources and should be used for non-investigative applications or in academic research only after the resources of other more traditional techniques have been thoroughly verified in their full range of possibilities [2,4,8].

Acknowledgments. We would like to thank the 30 volunteers who allowed their images and data to be acquired and disseminated for research purposes. This project was sponsored by FAPERJ (Rio de Janeiro State Research Support Foundation - Brazil) under grant number E-26/210.019/2020. A.C. received suport of CNPQ (Brazilian Agency).

References

1. Araujo, A.S., Issa, M.H.S., Sanchez, Á., Muchaluat-Saade, D.C., Conci, A.: Utilizing infrared thermography for assessing breast cancer treatment progression. In: 2023 30th International Conference on Systems, Signals and Image Processing (IWSSIP), pp. 1–5 (2023). https://doi.org/10.1109/IWSSIP58668.2023.10180273

2. Baffa, M., Conci, A.: Radiomics for breast IR-imaging classification. In: MICCAI Workshop on Artificial Intelligence over Infrared Images for Medical Applications (AIIIMA 2022). LNCS, vol. 13602, pp. 10–19. Springer, Cham (2022). https://doi.org/10.1007/978-3-031-19660-7_2

3. Baffa, M., Lattari, L., Conci, A.: Segmentation of Breast by 3D U-Nets. In: MICCAI 2023, 2nd Workshop on Artificial Intelligence over Infrared Images for Medical Applications (AIIIMA) (2023). https://doi.org/10.1007/978-3-031-44511-8

4. Baffa, M.F.O., Neves, T., Conci, A.: DCGANs for data augmentation in breast thermography. In: Kakileti, S.T., Manjunath, G., Schwartz, R.G., Ng, E.Y.K. (eds.) Artificial Intelligence over Infrared Images for Medical Applications (AIIIMA 2024). LNCS, vol. 15279, pp. 67–79. Springer, Cham (2024). https://doi.org/10.1007/978-3-031-76584-1_5

5. Crittenden, T., et al.: Measuring breast volume in hypertrophy: laser scanning or water displacement? Aust. J. Plast. Surg. **1**(2), 33–40 (2018). https://doi.org/10.34239/ajops.v1i2.119

6. Costa, G.M., Moura, E.L.S., Borchartt, T.B., Conci, A.: Modeling the 3D breast surface using thermography. In: Lecture Notes in Computer Science, pp. 45–56. Springer, Cham (2023). https://doi.org/10.1007/978-3-031-44511-8_3

7. Killaars, R.C., et al.: Clinical assessment of breast volume: can 3D imaging be the gold standard? Plast. Reconstr. Surg. Glob Open **8**(11), e3236 (2020). https://doi.org/10.1097/GOX.0000000000003236

8. Majidpour, J., Ahmed, H.A., Ahmed, M.H., et al.: Applications of GAN models in breast cancer detection: a comprehensive review. Arch. Comput. Methods Eng. (2025). https://doi.org/10.1007/s11831-025-10323-7

9. Mashekova, A., Zhao, Y., Ng, E.Y.K., Zarikas, V., Fok, S.C., Mukhmetov, O.: Early detection of the breast cancer using infrared technology – a comprehensive review. Thermal Sci. Eng. Progress (2022). https://doi.org/10.1016/j.tsep.2021.101142

10. Mayer, H.F., Coloccini, A., Vinas, J.V.: Three-dimensional printing in breast reconstruction: current and promising applications. J. Clin. Med. **13**(11), 3278 (2024). https://doi.org/10.3390/jcm13113278

11. Moura, E., Costa, G., Borchartt, T., Conci, A.: A tool for 3D representation of the 2D thermographic breast acquisitions. In: Proceedings of the 19th International Joint Conference on Computer Vision, Imaging and Computer Graphics Theory and Applications, Rome, Italy, p. 138 (2024). https://www.scitepress.org/Papers/2024/124695/124695.pdf

12. Moura, E.L.S., Conci, A.: Reconstrução Tridimensional a partir de Termografias: Avaliação de Modelos 3D e Simetria das Mamas. Simpósio Brasileiro de Computação Aplicada à Saúde, 2025, Porto Alegre/RS, pp. 103–108 (2025). ISSN 2763-8987. https://doi.org/10.5753/sbcas_estendido.2025.6844

13. Moura, E.S., Pastura, F.C.H., Scagliusi, N., Oliveira, W.I., Conci, A.: Breast symmetry classification: a proposal of image analysis based indices. In: LIQUE IMX 2025, Niterói, Brazil (2025). https://nca.ufma.br/lique2025

14. Rakhunde, M.B., Gotarkar, S., Choudhari, S.G.: Thermography as a breast cancer screening technique: a review article. Cureus **14**(11), e31251 (2022). https://doi.org/10.7759/cureus.31251. PMID: 36505165; PMCID: PMC9731505

15. Resmini, R., Silva, L., Araujo, A.S., Medeiros, P., Muchaluat-Saade, D., Conci, A.: Combining genetic algorithms and SVM for breast cancer diagnosis using infrared thermography. Sensors **21**, 4802 (2021). https://doi.org/10.3390/s21144802

16. Ryan, L., Agaian, S.: Breast cancer detection using infrared thermography: a survey of texture analysis and machine learning approaches. Bioengineering (Basel) **12**(6), 639 (2025). https://doi.org/10.3390/bioengineering12060639. PMID: 40564455; PMCID: PMC12189745

17. Sa, F., Conci, A.: On breast reconstruction using IR images by AI techniques. In: IMX Work in Progress Proceedings. SBC Open Library (2025). https://doi.org/10.1145/3706370.3731646

18. Saha, A.P., Kakileti, S.T., Dedhiya, R., Manhunath, G.: 3D-BreastNet: a self-supervised deep learning network for reconstruction of 3D breast surface from 2D thermal images. In: Artificial Intelligence over Infrared Images for Medical Applications. AIIIMA 2023. LNCS, vol. 14298. Springer, Cham (2023). https://doi.org/10.1007/978-3-031-44511-8_2

19. Santos, L.C., Lima, R.C.F., Paiva, A.C., Conci, A., Espindola, N.A.: A computing platform to analyze breast abnormalities using infrared images. Med. Biol. Eng. Comput. **61**, 305–315 (2023). https://doi.org/10.1007/s11517-022-02726-6

20. Silva, L.F., et al.: A new database for breast research with infrared image source. J. Med. Imaging Health Inf. **4**(1), 92–100(9) (2014). https://doi.org/10.1166/jmihi.2014.1226. E JMIHI ISSN: 2156-7018 (Print): EISSN: 2156-7026 (Online)

21. Veitch, D., Burford, K., Dench, P., DeanN, N., Griffin, P.: Measurement of breast volume using body scan technology (Computer-Aided Anthropometry). Work **41**(supl. 1), 4038–4045 (2012). https://doi.org/10.3233/WOR-2012-0068-4038. PMID: 22317340

MedThermal-DICOM: An Open-Source DICOM-Compliant Framework for Medical Thermal Imaging Enabling Clinical Integration and Research Reproducibility

Bharath Govindaraju, Siva Teja Kakileti, Ronak Dedhiya[✉], and Geetha Manjunath

Niramai Health Analytix Pvt Ltd., Bangalore, India
medthermaldicom@niramai.com

Abstract. Thermal imaging is emerging as a valuable diagnostic modality in medical applications, offering non-invasive assessment of physiological conditions through temperature mapping. However, the lack of standardized data formats and comprehensive software tools has severely limited clinical adoption and research reproducibility. Here we present MedThermal-DICOM, the first open-source Python framework that provides full compliance with the Digital Imaging and Communications in Medicine (DICOM) standard for thermal imaging. The framework preserves thermal-specific metadata through private extensions, ensures quantitative fidelity via Real World Value Mapping (RWVM), and integrates seamlessly with Picture Archiving and Communication Systems (PACS) and existing clinical workflows. Validation across three public datasets demonstrated syntactic and semantic compliance, quantitative accuracy better than 0.01 °C, and interoperability with both PACS and RWVM-aware software. By enabling standards-compliant thermal imaging with complete metadata preservation, MedThermal-DICOM provides the technical foundation for reproducible research, multi-center studies, and eventual clinical deployment. The framework is freely available under an open-source license to promote widespread adoption and collaborative development.

Keywords: Thermal Imaging · DICOM Standards · Medical Imaging · Python Framework · Clinical Workflow · Open Source · Thermal DICOM

1 Introduction

Thermal imaging has long been of interest in clinical medicine as a non-invasive, radiation-free modality capable of capturing physiological alterations through surface temperature distribution [1, 2]. Initial explorations in the 1960s highlighted its potential in oncology, particularly breast cancer detection, where tumors exhibit increased metabolic activity and angiogenesis that may manifest as localized thermal asymmetries or vascular patterns [3, 4]. However, interest declined immediately due to methodological limitations, variability in image acquisition and interpretation, and the absence of standardized protocols, preventing widespread clinical adoption.

© The Author(s), under exclusive license to Springer Nature Switzerland AG 2026
S. T. Kakileti et al. (Eds.): AIIIMA 2025, LNCS 16308, pp. 190–204, 2026.
https://doi.org/10.1007/978-3-032-10990-3_13

In recent years, advances in thermal camera hardware, infrared sensor technology, and machine learning have renewed interest in thermography as an adjunctive imaging tool [3]. Modern uncooled microbolometer-based cameras now offer sub-50 mK thermal sensitivity, higher spatial resolution, and improved calibration stability, addressing many of the technical limitations of earlier generations of equipment. These improvements, combined with algorithmic advances in machine learning, have enabled more reliable quantification of subtle thermal asymmetries and vascular heat patterns, thereby reinvigorating research into clinical applications of thermography [3–6].

Building on these advances, oncology has re-emerged as a primary focus of thermographic research [7–10]. AI-assisted breast cancer detection using thermography has shown promising diagnostic performance, with reported sensitivities and specificities approaching those of conventional imaging modalities [7, 8]. Beyond oncology, high-resolution thermal imaging with automated interpretation has been studied for predicting diabetic foot ulcers [11, 12] and monitoring wound healing [13], while also enabling assessment of circulatory disorders. In addition, thermography is being explored for fever screening [14], pain management [15], rehabilitation [16, 17], intraoperative perfusion monitoring, and emerging roles in liver disease [18], autism spectrum disorder [19], and neglected tropical diseases [20]. Together, these applications highlight how modern hardware and AI have transformed thermography into a versatile platform for detecting physiological changes linked to vascular, metabolic, and inflammatory processes.

Despite this promise, thermal imaging remains underutilized in mainstream workflows, largely due to the lack of standardized data representation. Most current systems store images in generic formats such as PNG or JPEG, which discard critical acquisition parameters including emissivity, object distance, ambient temperature, and calibration factors and lack support for patient demographics, study metadata, or longitudinal tracking. This results in poor reproducibility of AI algorithms, limited multi-center collaboration, and minimal integration with other imaging modalities.

Integration into Picture Archiving and Communication Systems (PACS) is particularly important as it enables secure storage, retrieval, and distribution of imaging studies while supporting longitudinal patient monitoring and multi-modality comparison [21]. The inability to archive and retrieve thermal imaging studies alongside modalities such as X-ray, CT, or MRI severely restricts clinical adoption. Without PACS integration, thermal imaging remains confined to standalone research settings rather than being embedded within the established diagnostic ecosystem.

The Digital Imaging and Communications in Medicine (DICOM) standard addresses this challenge. Since its introduction in the 1980s, DICOM has become the global standard for medical imaging interoperability, supporting nearly all conventional modalities including radiography, CT, MRI, ultrasound, PET, nuclear medicine, ophthalmic imaging, endoscopy, digital pathology, and visible light photography[22, 23]. The standard not only enables robust metadata encoding and secure data exchange but also ensures compatibility with PACS and clinical workflows. Extending DICOM support to thermal imaging is therefore critical to enable its systematic adoption in clinical practice, facilitate large-scale research, and ensure reproducibility across centers.

In this work, we present MedThermal-DICOM, an open-source Python library designed to generate DICOM-compliant files for thermal imaging. The framework introduces thermal-specific metadata fields, supports flexible UID and workflow management, allows temperature display, and provides full compatibility with PACS environments. By bridging the gap between thermography and the medical imaging ecosystem, MedThermal-DICOM establishes the foundation for standardized, interoperable use of thermal imaging in both clinical and research domains. We anticipate that this contribution will accelerate the clinical integration of thermography, foster reproducibility in research, and enable new opportunities for multi-center collaboration.

2 Methods

2.1 Framework Overview

MedThermal-DICOM was developed with three guiding principles: (i) compatibility with existing clinical imaging infrastructure, (ii) preservation of acquisition metadata to ensure reproducibility, and (iii) usability for both technical and non-technical users. To achieve these objectives, we implemented a modular, layered architecture (Fig. 1) in which each component addresses a distinct requirement while maintaining interoperability.

The foundation layer provides computational stability through Python (v3.8+), PyDICOM [24], and NumPy [25], enabling efficient array manipulation and standards-compliant file construction. The standards layer enforces DICOM compliance, validating identifiers, private block allocation, and header population. The core framework layer encodes radiometric pixel data, applies temperature transformations, and integrates both standard and thermal-specific metadata. Finally, the application layer delivers user-facing functionality: clinicians and researchers without programming expertise can generate DICOM-compliant thermal images through a graphical interface, while technical users can directly integrate the library into Python workflows.

This hierarchical design aligns each architectural element with an adoption outcome: stability, compliance, reproducibility, and usability.

2.2 Metadata Scheme

To ensure interoperability while preserving acquisition context, MedThermal-DICOM employs a two-level metadata strategy. At the first level, all mandatory DICOM modules such as Patient, Study, Series, and Image are populated in accordance with the DICOM Information Object Definitions. These modules include demographics, identifiers, timestamps, image geometry, and pixel data, thereby ensuring integration with standard PACS infrastructure. Anatomical context is explicitly represented using the Anatomic Region Sequence (0008,2218) populated with SNOMED CT terminology (for example, Breast [76752008] and Head [695360]) and, where appropriate, the Body Part Examined (0018,0015) attribute populated with DICOM-defined codes. This dual annotation guarantees compatibility across systems that may rely on either attribute for indexing. MedThermal-DICOM automatically establishes the corresponding Anatomic

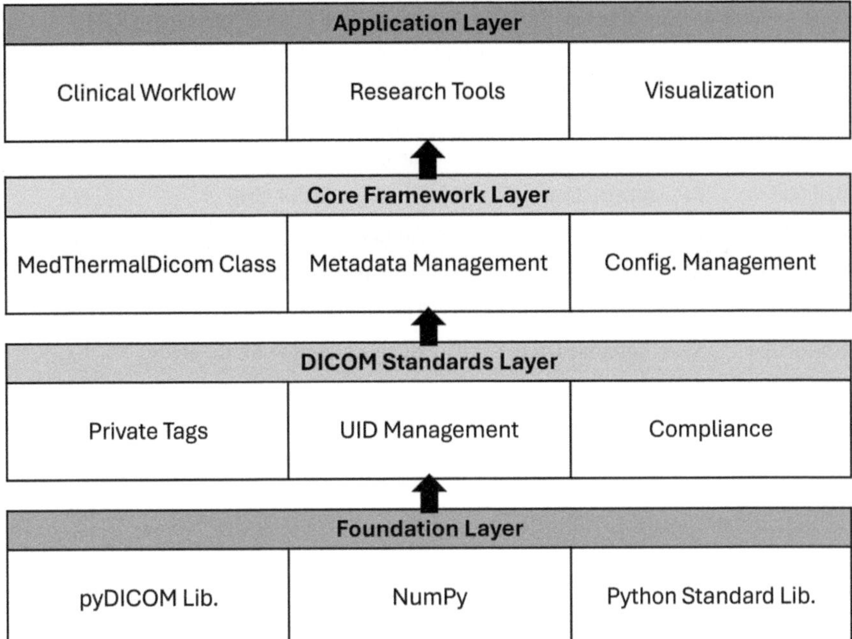

Fig. 1. MedThermal-DICOM Framework Architecture. The framework consists of modular components that provide distinct functionality while maintaining cohesive integration. Core components include the ThermalDicom class for DICOM dataset management, metadata management for standards compliance, utility functions for UID generation and validation, and configuration management for clinical and research environments.

Region Sequence and related codes when the Body Part Examined is set (illustrated Table 1).

At the second level, a structured private dictionary was introduced to capture parameters unique to thermal imaging not supported by the standard. 12 elements have been defined under the private creator "MedThermal-DICOM" within the group (0019,xxxx). These include emissivity, camera-to-subject distance, ambient and reflected temperatures, relative humidity, calibration date, detector sensitivity, device identifiers, and colormap settings. Each element is documented with its Value Representation (VR) and clinical significance (CS), ensuring that exported datasets retain the physical conditions necessary for reproducible thermal interpretation (illustrated Table 2).

By combining strict use of core DICOM fields with carefully designed private extensions, MedThermal-DICOM preserves both patient/study context and the physical conditions necessary for reproducible temperature interpretation. Further, the framework accommodates the use of other standard DICOM tags and user-defined private tags to meet one's requirements.

Table 1. Key Standard DICOM Attributes Used in MedThermal-DICOM.

Tag	Name	Description	VR
(0010,0010)	Patient Name	Full patient name used for identification	PN
(0010,0030)	Patient Birth Date	Patient's date of birth	DA
(0010,0040)	Patient Sex	Administrative sex of the patient	CS
(0008,0020)	Study Date	Date the study started	DA
(0008,0030)	Study Time	Time the study started	TM
(0020,000D)	Study Instance UID	Unique identifier for the study	UI
(0020,000E)	Series Instance UID	Unique identifier for the series	UI
(0020,0013)	Instance Number	Number identifying the image within the series	IS
(0008,0018)	SOPInstanceUID	Unique identifier for the image (instance)	UI
(0008,0016)	SOPClassUID	Identifies the type of DICOM object	UI
(0008,1030)	Study Description	Free-text description of the study	LO
(0008,2218)	Anatomic Region Sequence	Sequence identifying the anatomic region examined	SQ
(0008,0100)	Code Value	Code value representing the concept (from a coding scheme e.g., SNOMED CT)	SH
(0008,0102)	Coding Scheme Designator	Designator of the coding scheme (e.g., SCT = SNOMED CT)	SH
(0008,0104)	Code Meaning	Human-readable meaning of the code	LO
(0018,0015)	Body Part Examined	Free-text code string identifying the body part examined.	CS
(0028,0004)	Photometric Interpretation	Specifies the intended interpretation of the pixel data	CS
(0028,0010)	Rows	No. of rows in the image	US
(0028,0011)	Columns	No. of columns in the image	US
(0028,0030)	PixelSpacing	Physical distance between pixels, specified by row spacing and column spacing	DS
(0028,0100)	Bits Allocated	No. of bits allocated for each pixel sample	US
(0028,0101)	BitsStored	No. of bits used to store the pixel data	US

(continued)

Table 1. (*continued*)

Tag	Name	Description	VR
(0028,1101)	Red Palette Color LUT Descriptor	Defines the structure of the Red LUT (Number of Entries, First Pixel Mapped, Bits Per Entry)	SS/US
(0028,1102)	Green Palette Color LUT Descriptor	Defines the structure of the Green LUT	SS/US
(0028,1103)	Blue Palette Color LUT Descriptor	Defines the structure of the Blue LUT	SS/US
(0028,1201)	Red Palette Color LUT Data	Contains the list of Red intensity values That map to grayscale pixel values	OW
(0028,1202)	Green Palette Color LUT Data	Contains the list of Green intensity values That map to grayscale pixel values	OW
(0028,1203)	Blue Palette Color LUT Data	Contains the list of Blue intensity values That map to grayscale pixel values	OW
(2050,0020)	Presentation LUT Shape	Stores descriptive name for color palette/predefined LUT transformation	CS
(7FE0,0010)	Pixel Data	Encoded image pixel data	OW/OB
(0028,1052)	Rescale Intercept	Intercept for mapping stored values to real-world values	DS
(0028,1053)	Rescale Slope	Slope for mapping stored values to real-world values	DS
(0040,9096)	Real World Value Mapping Seq.	Defines mapping between stored values and °C measurements	SQ
(0008,0060)	Modality	Imaging modality (Thermography: 'TG')	CS
(0008,0070)	Manufacturer	Manufacturer of the camera (e.g., FLIR)	LO
(0008,1090)	Manufacturer's Model Name	Camera Model (e.g., E75)	LO
(0018,1000)	Device Serial Number	Camera serial number	LO
(0018,1200)	Date of Last Calibration	Calibration date	DA
(0018,9302)	Acquisition Type	Acquisition mode	CS
(0018,9424)	Acquisition Protocol Description	Thermal Imaging protocol description with environment conditions	LT

(continued)

Table 1. (*continued*)

Tag	Name	Description	VR
(0020,0060)	Laterality	Laterality of the examined body part	CS
(0018,5101)	ViewPosition	View of the image (e.g., AP, PA, LATERAL, OBLIQUE)	CS
(0020,0020)	PatientOrientation	Patient's orientation with respect to the image rows and columns (e.g., L\F)	CS
(0008,0090)	ReferringPhysicianName	Name of the physician who referred the patient for the study	PN

Table 2. Thermal-Specific DICOM Private Tags.

Tag	Name	Description	VR	Clinical Significance
(0019,1010)	Emissivity	Object emissivity coefficient	DS	Critical for temperature accuracy
(0019,1011)	Distance from Camera	Distance from camera to subject (meters)	DS	Required for calibration
(0019,1012)	Ambient Temperature	Ambient temperature	DS	Environmental correction
(0019,1013)	Reflected Temperature	Reflected temperature	DS	Atmospheric compensation
(0019,1014)	Atmospheric Temperature	Atmospheric temperature	DS	Environmental modeling
(0019,1015)	Relative Humidity	Relative humidity (%)	DS	Atmospheric effects
(0019,1016)	Temperature Range Min	Minimum temperature	DS	Scale calibration
(0019,1017)	Temperature Range Max	Maximum temperature	DS	Scale calibration
(0019,1018)	Temperature Unit	Temperature unit	LO	Unit specification
(0019,1019)	Thermal Sensitivity	NETD (mK)	DS	Performance metric
(0019,1020)	Spectral Range	Detector spectral range (μm)	LO	Technical specification
(0019,1021)	Lens Field of View	Optical field of view of the thermal camera lens, in degrees	DS	Important for spatial calibration and quantitative interpretation

2.3 Pixel Representation and Temperature Encoding

Thermal images are encoded as grayscale integer arrays with Photometric Interpretation = MONOCHROME2, preserving the bit depth of the acquisition device. Spatial descriptors such as Pixel Spacing (0028,0030), Image Orientation (0020,0037), and Image Position (0020,0032) are included when available, enabling alignment with other imaging modalities. Colormaps are stored as optional display metadata but do not alter the underlying grayscale pixel values. Quantitative temperature values are derived from the stored pixel data using the standard rescale operation.

$$T = stored\_value \times Rescale\ Slope\ (0028, 1053) + Rescale\ Intercept(0028, 1052). \tag{1}$$

To support explicit temperature representation, MedThermal-DICOM implements the Real World Value Mapping (RWVM) Sequence (0040,9096). Each dataset encodes slope, intercept, and valid temperature ranges. This allows RWVM-aware DICOM viewers to display per-pixel temperature values and region-of-interest statistics without reliance on proprietary software.

Colormaps are treated as optional display metadata. By default, pixel data are stored as quantitative grayscale values, and color thermal images are generated via lookup tables applied at display time. This preserves quantitative fidelity while allowing consistent rendering across different viewers.

2.4 Overlays and Derived Data

MedThermal-DICOM supports optional DICOM overlay planes (60xx group elements) for storing binary masks, annotations, or derived features alongside thermal images. This functionality allows integration of segmentation masks, ROI outlines, or analysis-derived bitmaps without altering the underlying pixel data. Overlays are encoded as separate bit-planes in compliance with the DICOM standard, ensuring that quantitative temperature values remain unchanged while interpretive layers are preserved for downstream display and analysis.

2.5 Dataset Organization and UID Management

MedThermal-DICOM adheres to standard DICOM conventions for organizing multi-view and multi-region thermal acquisitions. A single Study Instance UID (0020,000D) groups all images from the same exam session. Within a study, each anatomical region (e.g., breast, head, foot) is represented as a distinct Series, identified by a unique Series Instance UID (0020,000E). Individual views within a region are exported as sequential single-frame instances, ordered by Instance Number (0020,0013). This design mirrors MRI slice storage and ensures compatibility with PACS systems that natively support stack navigation.

Every image is uniquely identified by its SOP Instance UID (0008,0018), ensuring global traceability. UIDs are generated using a dual-mode strategy. In default mode (research/development), UIDs are created using PyDICOM utilities, which provide local uniqueness but are not guaranteed globally. In institutional mode (clinical/collaborative), organizations configure their own UID root, obtained from the DICOM Standards Committee (NEMA), from an institutional Object Identifier (OID) namespace, or via a national standards body. MedThermal-DICOM validates prefix assignment and prevents collisions. Both deterministic (reproducible from fixed inputs) and non-deterministic (randomized) UID generation are supported, depending on workflow requirements.

This organization ensures that multi-view, multi-region thermal datasets are fully queryable, uniquely identifiable, and seamlessly interoperable with PACS systems in both research and clinical environments.

Dataset assembly is managed by the MedThermal-DICOM class, which integrates radiometric arrays, standard modules, private metadata, overlays, and RWVM definitions into a complete DICOM object.

2.6 SOP Classes and Transfer Syntax

Thermal datasets are exported using Secondary Capture Image Storage (1.2.840.10008.5.1.4.1.1.7) as the SOP Class. The Modality attribute (0008,0060) is set to "TG" (Thermography), a designation supported in PACS environments and standardized in the DICOM registry. MedThermal-DICOM adopts this convention while noting the need for an official thermal modality code.

By default, files are written using uncompressed Implicit VR Little Endian and Explicit VR Little Endian transfer syntaxes. Lossy compression is not supported to avoid corruption of quantitative values. Optional lossless JPEG-LS and JPEG2000 compression are implemented for use cases requiring reduced storage, but are disabled by default to maximize interoperability.

2.7 Graphical User Interface

Although MedThermal-DICOM is fully accessible via its Python API, a graphical user interface (shown in Fig. 2) was developed to support users without programming expertise. The interface provides dedicated panels for entering patient and study metadata, specifying acquisition parameters such as emissivity, distance, and environmental conditions and selecting export options including UID roots. All GUI operations call the same programmatic API as the command-line and scripting workflows, ensuring consistency of outputs across use modes.

2.8 PACS Integration and Workflow Support

Thermal DICOM objects generated by MedThermal-DICOM were validated for interoperability with existing PACS systems. Files were imported via file-system ingestion and successfully archived, indexed, and retrieved using C-FIND and C-MOVE. RWVM-enabled viewers displayed correct quantitative temperatures, and private metadata remained accessible through inspection tools.

Although the current release relies on file-based exchange, a C-STORE SCU utility is provided for direct network transmission, and DICOMweb (STOW-RS/QIDO-RS/WADO-RS) support is planned.

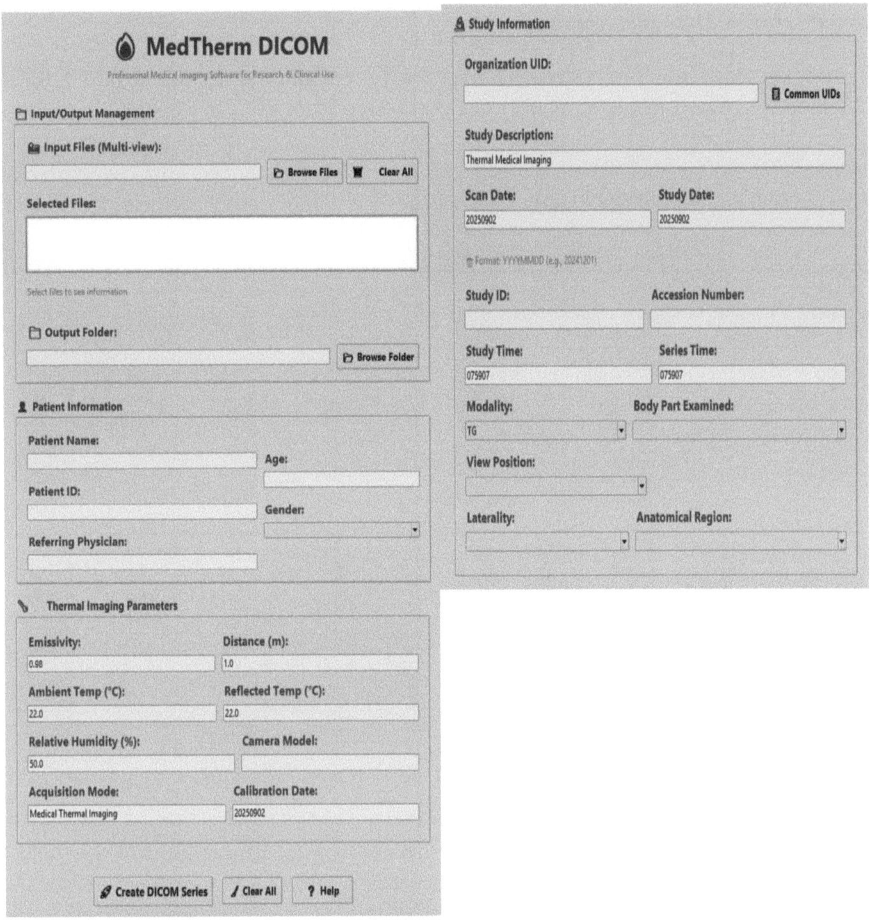

MedTherm DICOM
Professional Medical Imaging Software for Research & Clinical Use

Input/Output Management

Input Files (Multi-view):
Browse Files   Clear All

Selected Files:

Select files to see information

Output Folder:
Browse Folder

Patient Information

Patient Name:
Age:
Patient ID:
Gender:
Referring Physician:

Thermal Imaging Parameters

Emissivity: 0.98
Distance (m): 1.0
Ambient Temp (°C): 22.0
Reflected Temp (°C): 22.0
Relative Humidity (%): 50.0
Camera Model:
Acquisition Mode: Medical Thermal Imaging
Calibration Date: 20250902

Create DICOM Series   Clear All   Help

Study Information

Organization UID:
Common UIDs

Study Description:
Thermal Medical Imaging

Scan Date: 20250902
Study Date: 20250902

Format: YYYYMMDD (e.g., 20241201)

Study ID:
Accession Number:

Study Time: 075907
Series Time: 075907

Modality: TG
Body Part Examined:

View Position:

Laterality:
Anatomical Region:

Fig. 2. Graphical user interface (GUI) of MedThermal-DICOM.

3   Results

3.1   Installation and Open-Source Demonstration

MedThermal-DICOM is openly released under the MIT License and can be installed from source, enabling reproducibility across sites. To demonstrate generalizability across diverse acquisition contexts, we applied the framework to four publicly available datasets: the DMR-IR Breast Thermography dataset [26], the Standup Thermogram Database for Diabetic Foot Complications [27], Thyroid Nodule thermal imaging dataset [28] and the Onchocerciasis Thermal Imaging dataset [20]. These datasets span distinct anatomical regions, camera models, and acquisition conditions, with image resolutions ranging from 160×120 to 640×480 pixels and temperature ranges between 20 and 40 °C. Together, they provided a representative benchmark for evaluating compliance, fidelity, interoperability, and performance.

3.2 Workflow Demonstration

A minimal workflow for converting CSV-based plantar thermograms into DICOM objects is illustrated in the Appendix Section (Algorithm 1). The generated files were successfully imported into Weasis and RadiAnt viewers (Fig. 3), where standard cursor readouts and ROI tools displayed quantitative temperatures in °C. This confirmed that exported objects were immediately interpretable in widely used viewing environments.

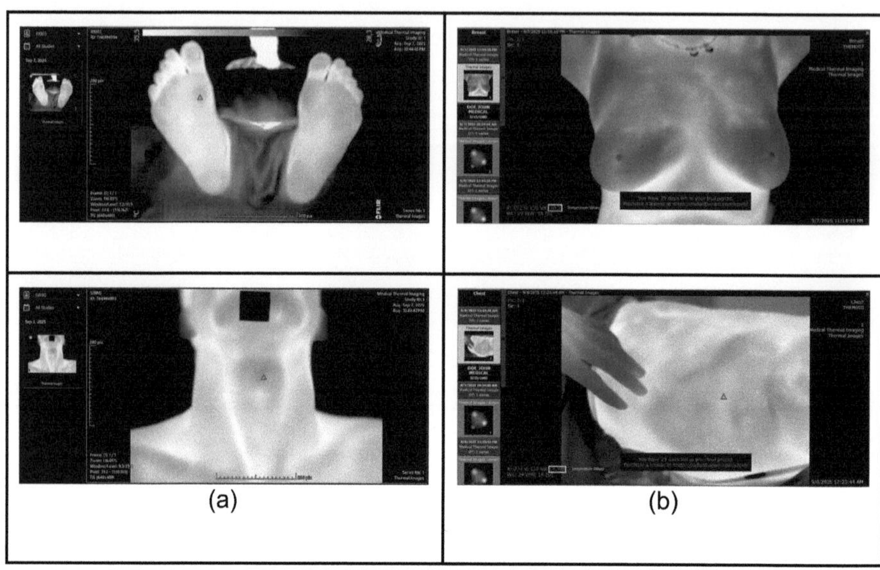

Fig. 3. (a) Example thermal DICOM loaded in Weasis viewer with quantitative temperature readouts. (b) Example thermal DICOM loaded in Radiant viewer with quantitative temperature readouts.

3.3 Standards Compliance and Metadata Integrity

All exported objects passed syntactic and semantic validation using PyDICOM and the DCMTK tool suite. Mandatory DICOM modules—including Patient, Study, Series, and Image—were consistently present, and no malformed attributes were detected. Private elements defined in the MedThermal-DICOM dictionary were preserved across exports, ensuring that acquisition parameters such as emissivity, camera distance, and ambient conditions remained available for downstream analysis.

3.4 Quantitative Temperature Fidelity

Fidelity of temperature encoding was assessed through round-trip conversion of synthetic and clinical datasets. Synthetic arrays with known values were exported and re-ingested without loss, yielding exact reconstruction of pixel intensities and derived temperatures. Clinical calibration images further confirmed the correct population of the RWVM sequence: per-pixel temperatures derived from slope and intercept were indistinguishable from reference values, with mean absolute error consistently below 0.01 °C—a residual attributable primarily to quantization effects from digitizing the temperature values. These findings demonstrate that MedThermal-DICOM preserves radiometric accuracy required for quantitative interpretation.

3.5 Interoperability with PACS and Viewers

Exported datasets were archived and retrieved through an open-source Dicoogle PACS (illustrated in Fig. 4). All C-STORE, C-FIND, and C-MOVE operations completed without metadata loss. RWVM-aware viewers, including Weasis and 3D Slicer, displayed per-pixel temperatures directly from the exported objects. Private thermal tags, while not rendered by these viewers, remained accessible in inspection tools, confirming end-to-end preservation of extended metadata. Together, these results demonstrate seamless integration of MedThermal-DICOM into existing PACS workflows and open-source viewers.

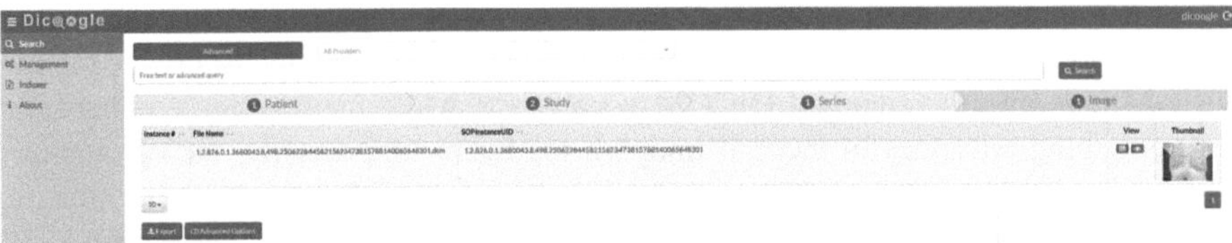

Fig. 4. Screenshot of Dicoogle PACS after archiving MedThermal-DICOM datasets, confirming query, retrieval, and metadata preservation.

4 Discussion

MedThermal-DICOM advances the field of thermal imaging by providing the first open-source framework that encodes radiometric data in full compliance with the DICOM standard. In contrast to the prevailing reliance on proprietary or generic formats (e.g., JPEG), which often discard acquisition metadata, the framework preserves critical parameters such as emissivity, camera-to-object distance, and environmental conditions as private tags within the DICOM structure. This ensures that thermal data are retained in a clinically interpretable form and remain reproducible across institutions, overcoming a major obstacle to wider adoption of thermal imaging.

The framework also demonstrates seamless interoperability with PACS and DICOM viewers, addressing a long-standing barrier that has confined thermography to research contexts. By supporting both file-based ingestion and network-based exchange, MedThermal-DICOM aligns thermal imaging with the same workflow conventions as CT, MRI, and ultrasound. Such integration enables not only secure archiving and retrieval but also longitudinal monitoring and multi-modality comparison, features essential for patient management in oncology, diabetic care, and surgical applications.

Equally important is the emphasis on quantitative fidelity. Through Real World Value Mapping, MedThermal-DICOM allows pixel-level temperatures to be reconstructed and displayed in RWVM-aware viewers without reliance on proprietary software. Validation on diverse datasets confirmed sub-0.01 °C error due to this inherent quantization, underscoring its suitability for clinical and research applications that depend on precise thermal quantification.

From a research perspective, standardized encoding of both images and metadata facilitates reproducibility, secondary analyses, and AI development. Machine learning models for breast cancer detection, diabetic foot ulcer prediction, and perfusion monitoring can now be trained on datasets that retain full acquisition context, enhancing generalizability and clinical trust. Furthermore, the availability of a common encoding framework lowers the barrier to multi-center collaborations, a key requirement for building robust evidence around the clinical utility of thermography. Early proposals demonstrated the feasibility of adopting DICOM for thermal imaging nearly two decades ago [29], but progress since then has been limited. By providing a comprehensive and open-source implementation, MedThermal-DICOM extends this foundational vision into a practical and scalable framework for contemporary use.

Despite these advances, several areas remain for future development. The current implementation does not yet include automated calibration workflows or AI-based analysis pipelines, both of which are increasingly important for clinical decision support. Future iterations will also explore cloud-based processing, standardized calibration algorithms, and collaboration with the DICOM community to incorporate the private thermal DICOM tags into the standard DICOM vocabulary, while maintaining compatibility with emerging advances in thermal imaging hardware. Importantly, the open-source design ensures that such enhancements will remain accessible to the broader imaging community, fostering community-driven development and accelerating adoption.

5   Conclusion

MedThermal-DICOM provides the first open-source, DICOM-compliant framework for thermal imaging, addressing long-standing challenges in standardization, interoperability, and metadata preservation. By ensuring seamless integration with PACS and clinical workflows, it enables reproducible research, multi-center collaboration, and quantitative analysis alongside established modalities such as CT, MRI, and ultrasound. This framework lays the foundation for thermography's transition from a niche research tool to a clinically robust imaging modality, while also creating a platform for future AI-driven applications.

Code Availability Statement. The complete source code for ThermalDICOM is available at: https://github.com/medthermaldicom/MedThermalDicom. The framework is distributed under the MIT License, enabling unrestricted use, modification, and distribution. Installation instructions and usage examples are provided in the repository documentation.

References

1. Ring, E.F., Ammer, K.: Infrared thermal imaging in medicine. Physiol. Meas. **33**(3), R33 (2012)
2. Lahiri, B.B., Bagavathiappan, S., Jayakumar, T., Philip, J.: Medical applications of infrared thermography: a review. Infrared Phys. Technol. **55**(4), 221–235 (2012)

3. Kakileti, S.T., Manjunath, G., Madhu, H., Ramprakash, H.V.: Advances in breast thermography. New Perspect. Breast Imaging **4**, 91–103 (2017)
4. Kennedy, D.A., Lee, T., Seely, D.: A comparative review of thermography as a breast cancer screening technique. Integr. Cancer Ther. **8**(1), 9–16 (2009)
5. Liu, Q., et al.: Infrared thermography in clinical practice: a literature review. Eur. J. Med. Res. **30**(1), 33 (2025)
6. Kakileti, S.T., Manjunath, G.: AIM for breast thermography. In: Artificial Intelligence in Medicine, pp. 1–16. Springer International Publishing, Cham (2021)
7. Singh, D., Singh, A.K.: Role of image thermography in early breast cancer detection – past, present and future. Comput. Methods Programs Biomed. **1**(183), 105074 (2020)
8. Shrivastava, R., Kakileti, S.T., Manjunath, G.: Thermal radiomics for improving the interpretability of breast cancer detection from thermal images. In: MICCAI Workshop on Medical Image Assisted Blomarkers' Discovery, pp. 3–9. Springer Nature Switzerland, Cham (2022)
9. Etehadtavakol, M., Etehadtavakol, M., Moallem, G., Ng, E.Y.: Evaluating radiomics feature reduction for thyroid nodule segmentation in thermal imaging. In: MICCAI Workshop on Artificial Intelligence Over Infrared Images for Medical Applications, pp. 69–87. Springer Nature Switzerland, Cham (2024)
10. Magalhaes, C., Tavares, J.M., Mendes, J., Vardasca, R.: Comparison of machine learning strategies for infrared thermography of skin cancer. Biomed. Signal Process. Control **1**(69), 102872 (2021)
11. Dedhiya, R., Prasad, R.V., Kakileti, S.T., Manjunath, G.: Thermal radiomics for early detection of diabetic foot ulcers using infrared thermography. In: MICCAI Workshop on Artificial Intelligence Over Infrared Images for Medical Applications, pp. 1–10. Springer Nature Switzerland, Cham (2024)
12. Alevizos, V., Arampidis, N., Boja, I., Papakostas, G.A.: Reverse circular logarithmic LBP for diabetic foot ulcer detection. In: MICCAI Workshop on Artificial Intelligence over Infrared Images for Medical Applications, pp. 11–22. Springer Nature Switzerland, Cham (2024)
13. Ramirez-GarciaLuna, J.L., Bartlett, R., Arriaga-Caballero, J.E., Fraser, R.D., Saiko, G.: Infrared thermography in wound care, surgery, and sports medicine: a review. Front. Physiol. **3**(13), 838528 (2022)
14. Katte, P., Kakileti, S.T., Madhu, H.J., Manjunath, G.: Automated thermal screening for COVID-19 using machine learning. In: MICCAI Workshop on Medical Image Assisted Blomarkers' Discovery, pp. 73–82. Springer Nature Switzerland, Cham (2022)
15. Liu, H., Zhu, Z., Jin, X., Huang, P.: The diagnostic accuracy of infrared thermography in lumbosacral radicular pain: a prospective study. J. Orthop. Surg. Res. **19**(1), 409 (2024)
16. Cruz-Albarran, I.A.: Assessment of range of motion before and after hamstring percussion therapy using thermography and CNN. In: Artificial Intelligence over Infrared Images for Medical Applications: Third International Conference, AIIIMA 2024, Virtual Event, Proceedings, vol. 15279, p. 88. Springer Nature (2024)
17. Ahalya, R.K., Snekhalatha, U.: CNN transformer for the automated detection of rheumatoid arthritis in hand thermal. In: Artificial Intelligence over Infrared Images for Medical Applications: Third International Conference, AIIIMA 2024, Virtual Event, Proceedings, vol. 15279, p. 23. Springer Nature (2024)
18. Lan, Q., Sun, H., Robertson, J., Deng, X., Jin, R.: Non-invasive assessment of liver quality in transplantation based on thermal imaging analysis. Comput. Methods Programs Biomed. **1**(164), 31–47 (2018)
19. Ganesh, K., Umapathy, S., Thanaraj, K.P.: Deep learning techniques for automated detection of autism spectrum disorder based on thermal imaging. Proc. Inst. Mech. Eng. H **235**(10), 1113–1127 (2021)

20. Dedhiya, R., et al.: Evaluation of non-invasive thermal imaging for detection of viability of onchocerciasis worms. In: 2022 44th Annual International Conference of the IEEE Engineering in Medicine & Biology Society (EMBC), pp. 3518–3521. IEEE (2022)
21. Clark, T.: PACS and imaging informatics: basic principles and applications. Biomed. Instrum. Technol. **40**(2), 125 (2006)
22. Pianykh, O.S.: Digital Imaging and Communications in Medicine (DICOM): A Practical Introduction and Survival Guide. Springer Berlin Heidelberg, Berlin, Heidelberg (2012)
23. Mustra, M., Delac, K., Grgic, M.: Overview of the DICOM standard. In: 2008 50th International Symposium ELMAR, vol. 1, pp. 39–44. IEEE (2008)
24. Mason, D.L., et al.: pydicom: An Open Source DICOM Library. https://github.com/pydicom/pydicom. Accessed 06 September 2025
25. Harris, C.R., et al.: Array programming with NumPy. Nature **585**(7825), 357–362 (2020)
26. Silva, L.F., et al.: A new database for breast research with infrared image. J. Med. Imaging Health Inform. **4**(1), 92–100 (2014)
27. Bouallal, D., et al.: STANDUP database of plantar foot thermal and RGB images for early ulcer detection. Open Res. Europe **2**, 77 (2022)
28. González, J.R., Damião, C., Conci, A.: An infrared thermal images database and a new technique for thyroid nodules analysis. In: MEDINFO 2017: Precision Healthcare through Informatics, pp. 384–387. IOS Press (2017)
29. Schaefer, G., Huguet, J., Zhu, S.Y., Plassmann, P., Ring, F.: Adopting the DICOM standard for medical infrared images. In: 2006 International Conference of the IEEE Engineering in Medicine and Biology Society, pp. 236–239. IEEE (2006)

Author Index